Wilfred W. J. Hulsbergen

Conjectures in Arithmetic Algebraic Geometry

Aspects of Mathematics

Edited by Klas Diederich

*A Publication of the Max-Planck-Institut für Mathematik, Bonn

Volumes of the German-language subseries "Aspekte der Mathematik" are listed at the end of the book.

Wilfred W. J. Hulsbergen

Conjectures in Arithmetic Algebraic Geometry

A Survey

vieweg

Wilfred W. J. Hulsbergen
KMA, NL-4800 RG Breda
The Netherlands

Die Deutsche Bibliothek - CIP-Einheitsaufnahme

Hulsbergen, Wilfred W. J.:
Conjectures in arithmetic algebraic geometry: a survey /
Wilfred W. J. Hulsbergen. - Braunschweig; Wiesbaden:
Vieweg, 1992
(Aspects of mathematics: E; Vol. 18)

NE: Aspects of mathematics / E

AMS subject classification: 11G, 11M, 14C, 14G, 14H, 14K, 19D, 19E, 19F.

Softcover reprint of the hardcover 1st edition 1992

Vieweg is a subsidiary company of the Bertelsmann Publishing Group International.

Cover design: Wolfgang Nieger, Wiesbaden
Printed on acid-free paper

ISSN 0179-2156
ISBN 978-3-528-06433-4 ISBN 978-3-322-85466-7 (eBook)
DOI 10.1007/ 978-3-322-85466-7

Contents

Introduction

In this expository paper we sketch some interrelations between several famous conjectures in number theory and algebraic geometry that have intrigued mathematicians for a long period of time.

Starting from Fermat's Last Theorem one is naturally led to introduce L-functions, the main motivation being the calculation of class numbers. In particular, Kummer showed that the class numbers of cyclotomic fields play a decisive role in the corroboration of Fermat's Last Theorem for a large class of exponents. Before Kummer, Dirichlet had already successfully applied his L-functions to the proof of the theorem on arithmetic progressions. Another prominent appearance of an L-function is Riemann's paper where the now famous Riemann Hypothesis was stated. In short, nineteenth century number theory showed that much, if not all, of number theory is reflected by properties of L-functions.

Twentieth century number theory, class field theory and algebraic geometry only strengthen the nineteenth century number theorists's view. We just mention the work of E. Hecke, E. Artin, A. Weil and A. Grothendieck with his collaborators. Hecke generalized Dirichlet's L-functions to obtain results on the distribution of primes in number fields. Artin introduced his L-functions as a non-abelian generalization of Dirichlet's L-functions with a generalization of class field theory to non-abelian Galois extensions of number fields in mind. Weil introduced his zeta-function for varieties over finite fields in relation to a problem in number theory. Finally, with the invention of ℓ-adic cohomology by Grothendieck, all of the L-functions mentioned above could be incorporated in a framework which comprises classical number theory as part of modern algebraic geometry. In terms of schemes a number field is just a zero-dimensional object, but it is already highly non-trivial, e.g. its étale cohomology can be identified with its Galois cohomology which plays a very important role in class field theory.

For higher dimensional varieties over number fields one has a representation of the Galois group of the number field on the ℓ-adic cohomology of those varieties. The inverses of the characteristic polynomials of the (geometric) Frobenius element of the Galois group acting on the ℓ-adic cohomology of the various reductions of the variety are now the building blocks of the L-function of the variety. Unfortunately there are many more conjectures than proven theorems on these L-functions. All the conjectures discussed in this book will be about the behaviour of the L-functions at special integral values of their arguments, e.g. the order of a zero or a pole and the value of the first non-zero coefficient of the Taylor series of the L-function. In a sense these conjectures are suggested by results or other conjectures in the lowest dimensional cases of number fields and elliptic curves.

Finally one may 'break a variety into pieces' and define L-functions for the associated motives. As the theory of motives is still not completely satisfactory we tried to postpone as long as possible any reference to it. So we have stated most conjectures and results only for varieties, and for simplicity of the statements the base field will often be the field of rational numbers. In later chapters, however, we can not avoid the more general situation and then we shall freely use the language of (Deligne) motives and other base fields. For convenience a few pages on these motives are included as they play an increasing role in the literature and, of course, because they are a very useful instrument. However, no crystalline aspects are discussed.

The first two chapters are an expanded version of a lecture given at the Royal Military Academy at Breda, the Netherlands, for an audience of mostly applied mathematicians and computer scientists. The main purpose of the lecture was to demonstrate the use of several mathematical programs for a personal computer in the calculation of class numbers and rational points on elliptic curves. These programs are used mainly for educational purposes and we wanted to demonstrate that their use may lead to numerical results for more advanced problems. The calculations are not included here as they do no justice to the more advanced level of the final version of the text. Anyhow, no new numerical results were obtained.

As the original title of the lecture also refered to general L-functions and Beilinson's conjectures, which for lack of time we could not even mention then, we were asked to edit a written version of the text. The present paper was formulated out of this suggestion. As a quick look at its contents may reveal, it seemed useful to us to give a lengthy introduction (Chapters 1, 2, 3 and 4), without making the paper completely self-contained. Also, on several places we have anticipated concepts that are defined only in a later chapter where their precise description is more important. In particular, the notion of motive pervades the text from an early stage, while it is defined only in Chapter 8.

To give full proofs of all theorems would have made the book unwieldily lengthy. In particular, each of the introductury first four chapters contains already enough material to write a book. Besides, many easily accessible references are given. So, in these introductory chapters there are no proofs at all. In the later chapters we indicate the line of thought that eventually leads to a theorem, and again, the reader should consult the references in the bibliography to find full proofs. It should be mentioned (and will be evident) that we borrowed a lot from most of the contributions in the beautiful book [RSS], the expository paper [Ra3], and, with respect to the Hodge and Tate conjectures, [Ja2]. We accept full responsibility for any error or misinterpretation of their results.

As to the contents of the remaining chapters we add a few words: In Chapter 5 Beilinson's first conjecture is motivated by means of Borel's regulator map, which suggests to take algebraic K-theory into account, and of Deligne's conjecture to relate the regulator to the values of L-functions, the unifying object being Deligne-Beilinson cohomology. This is further motivated by Bloch's regulator map for curves. The words K-theory, regulator and Deligne cohomology will remain the key words for the rest of the text. In Beilinson's philosophy K-theory accounts for a universal cohomology theory and the regulator map represents the Chern character to any other suitable cohomology theory. It defines a $\mathbf{Q}$-structure on such a cohomology whose volume should be equal to the first non-zero coefficient of the L-function under consideration. In Chapter 6 these ideas are extended to include algebraic cycles and a conjecture of Tate. As an application the case of a Hilbert modular surface is discussed in some detail. Here, as well as in other known cases, a general phenomenon occurs: only part of the K-theoretic cohomology is necessary to obtain the sought for $\mathbf{Q}$-structure. In Chapter 7 we follow Gillet & Soulé to introduce a higher dimensional Arakelov intersection theory for arithmetic varieties. Beilinson's third conjecture is stated. This conjecture generalizes the Birch & Swinnerton-Dyer conjectures. In Chapters 8 and 9 an interrelation between the Hodge and Tate conjectures (also for singular varieties) is discussed. The existence of a Poincaré duality theory as explained in Chapter 3 is essential. A survey of Beilinson's construction of absolute Hodge cohomology, which generalizes Deligne-Beilinson cohomology, is given, as well as Jannsen's results on the relation between the regulator map and the Abel-Jacobi map. The introduction of motives and tannakian categories becomes natural and indispensable. In Chapter 9 mixed realizations and the tannakian subcategory of mixed motives are introduced. This may be the starting point for future developments such as Deligne's theory of a motivic fundamental group. Chapters 8 and 9 are inspired very much by Jannsen's work. In Chapter 10 a few examples are given, providing evidence for the truth of some of the conjectures stated in previous chapters.

Acknowledgements. I would like to thank J. van de Craats, G. van der Geer and J. Top for useful comments and critical remarks. I am especially grateful to J. Stienstra for pointing out several inaccuracies and many misprints in the original version of the manuscript.

Chapter 1

The zero-dimensional case: number fields

In this chapter two apparently different kinds of L-functions are introduced: Dirichlet and Artin L-functions. The main motivation is Dedekind's Class Number Formula, one of the highlights of nineteenth century number theory. This formula contains, among other things, an important entity, the regulator. This regulator and its generalizations will play a fundamental role in some of the most intriguing conjectures on L-functions of recent times. These conjectures, due to A. Beilinson, will be discussed in later chapters.

In spite of many proven results in number theory, e.g. the class number formula, there remain several deep conjectures even in this zero-dimensional case: Fermat's Last Theorem, the Riemann Hypothesis and Artin's Conjecture, to name just a few. Their connections with L-functions is the subject of this chapter and the next one.

1.1 Class Numbers

The famous Fermat Conjecture or Fermat's Last Theorem, which dates from about 1635, states:

Conjecture 1.1.1 (Fermat) *The equation* $x^l + y^l = z^l$, $l = 3, 4, 5, \cdots$, *has no non-trivial solutions in integers* x, y, z.

Until this moment it has remained a challenge either to prove this conjecture or find a counter-example.

Fermat, using his method of infinite descent, gave a proof for the case $l = 4$ and, as everybody knows, announced, in the margin ... etc.

About 1774 Euler gave a proof for $l = 3$, (actually Legendre gave the finishing touch to Euler's proof, which was incomplete), then (1825) Dirichlet and Legendre proved the case $l = 5$, Dirichlet proved $l = 14$ in 1832 and Lamé succeeded in proving the case $l = 7$. So far till about 1840, (cf. [No]).

One immediately sees that it suffices to restrict to prime exponents l, to prove the general conjecture. For powers of the prime 2 one uses Fermat's result.

The proofs in the above cases are rather special and no general method of attack to the problem was known until Ernst Eduard Kummer (1810–1893) started working on it. No doubt, Kummer was inspired by a proof for the case $l = 3$ that had been given by Gauß, who, however, never paid special attention to Fermat's Last Theorem.

As a matter of fact Gauß factorized the expression $x^3 + y^3$ into a product $(x+y)(x+\zeta y)(x+\zeta^2 y)$, where ζ is a primitive third root of unity, and considered the factors of this product as 'integers' in the 'number field $\mathbf{Q}(\zeta)$', consisting of (complex) numbers of the form $a + b\zeta + c\zeta^2$, a, b, $c \in \mathbf{Q}$, with the obvious addition and multiplication. In other words, Gauß constructed the so-called **cyclotomic field** $\mathbf{Q}(\zeta)$ with its **ring of integers** $\mathbf{Z}[\zeta]$.

Gauß's reasoning went as follows: if (x, y, z) were a non-trivial solution of the equation $x^3 + y^3 = z^3$, then each factor $x + y$, $x + \zeta y$ and $x + \zeta^2 y$ would be a third power (up to a unit) in the ring $\mathbf{Z}[\zeta]$. With a certain amount of work one can show that this leads to a contradiction. A warning is necessary: in this reasoning one needs **unique factorization** in $\mathbf{Z}[\zeta]$. Of course, Gauß was well aware of this: the ring $\mathbf{Z}[\zeta]$ is indeed a unique factorization domain. But in general unique factorization fails to hold. As an example, take the **imaginary field** $\mathbf{Q}(\sqrt{-5})$, obtained by adjoining $\sqrt{-5}$ to the rational numbers $\mathbf{Q}$, and its ring of integers $\mathbf{Z}[\sqrt{-5}]$, then 3, 7, $1+2\sqrt{-5}$ and $1-2\sqrt{-5}$ are primes, but 21 has two different factorizations: $21 = 3.7 = (1+2\sqrt{-5})(1-2\sqrt{-5})$. This phenomenon occurs in many number fields, especially in the higher cyclotomic fields $\mathbf{Q}(\zeta_m)$, where ζ_m is a primitive m^{th} root of unity and

$m \geq 20$. This makes the obvious generalization of Gauß's method, i.e. to write $x^l + y^l$ as a product $(x+y)(x+\zeta_l y)\cdots(x+\zeta_l^{l-1}y)$ and to use the same reasoning in $\mathbf{Z}[\zeta_l]$ as above, inappropriate to prove Fermat's Last Theorem in general.

A measure for the non-uniqueness of factorization in the ring of integers of a number field is given by the **class number** h of that number field. This is a positive integer, equal to one if and only if there is unique factorization. Thus one has e.g. $h(\mathbf{Q}(\zeta_3)) = h(\mathbf{Q}(\sqrt{-3})) = 1$, giving unique factorization and hence an affirmative answer to Fermat's Conjecture in case $l = 3$, but $h(\mathbf{Q}(\sqrt{-5})) = 2$ and $h(\mathbf{Q}(\sqrt{-23})) = 3$ ...etc. In particular, for the cyclotomic fields one has $h(\mathbf{Q}(\zeta_l)) = 1$ for $l = 3, \ldots, 19$, but $h(\mathbf{Q}(\zeta_{37})) = 37$ and $h(\mathbf{Q}(\zeta_{97})) = 577 \cdot 3457 \cdot 206209$, ..., etc.

For an arbitrary number field K with ring of integers $\mathcal{O}_K$ the **class number** is defined as the order of the **ideal class group** $Cl(\mathcal{O}_K)$, i.e. the abelian group of fractional ideals of $\mathcal{O}_K$ in K modulo the group of non-zero principal ideals. A basic result of classical number theory is the following

Theorem 1.1.1 *The class number $h(K) = \#Cl(\mathcal{O}_K)$ is finite.*

In general it is very difficult to calculate the class number, though for imaginary or real quadratic fields or cyclotomic fields (rather easy) expressions for the class number can be derived, cf. Section 1.4.

A remark on Kummer's work is appropriate. Kummer made a deep study of cyclotomic fields and by refined methods succeeded in proving many results on Fermat's Conjecture. Maybe the most impressive result is the following theorem:

Theorem 1.1.2 (Kummer(1847)) *Fermat's Conjecture is true for those prime exponents l which do not divide the class number of the cyclotomic field $\mathbf{Q}(\zeta_l)$.*

Such primes are called **regular**, and those that are not regular are called **irregular**. For example, 37 is an irregular prime and actually of all primes less than 100 only 37, 59 and 67 are irregular. In Kummer's days numerical evidence pointed to the belief that there are (far) more regular than irregular primes, but surprising as it may be, one can prove that there exist infinitely many irregular primes, whereas the question remains open as to the number of regular primes.

That it may be really difficult to prove specific statements on class numbers can be illustrated by the following example, which is known in the literature as Gauß's Class Number Problem for Imaginary Quadratic Fields:

Conjecture 1.1.2 (Gauß) *The class number $h(\mathbf{Q}(\sqrt{d}))$ of the imaginary quadratic field $\mathbf{Q}(\sqrt{d})$, $d < 0$ and square free, equals* one *if and only if* $d = -1, -2, -3, -7, -11, -19, -43, -67,$ *or* -163.

The conjecture gave rise to much research by later generations of mathematicians, but it was not until 1967 that it was proved by H. Stark. In the same spirit A. Baker and, independently, H. Stark proved in 1971 that there are exactly eighteen imaginary quadratic fields with class number 2. The ultimate result which solves (up to a finite amount of computation) the general Gauß Class Number Problem is the following theorem due to D. Goldfeld, B. Gross and D. Zagier.

Theorem 1.1.3 *For every $\varepsilon > 0$ there is an effectively computable constant $c > 0$ such that* $h(\mathbf{Q}(\sqrt{d})) > c(\log|d|)^{1-\varepsilon}$.

The proof depends on a deep result of Gross and Zagier from the theory of so-called 'Heegner points' on modular curves. For details we refer to Ch.10, [Go] and [GZ].

1.2 Dirichlet L-Functions

The search for a formula or a method to calculate the class number became an evident task of number theorists in the nineteenth century, already under the influence of Gauß with his study of quadratic forms and a little later inspired by Kummer's work on cyclotomic fields in relation to Fermat's Last Theorem. Dirichlet developed a formula for quadratic fields, and he and Kummer found one for cyclotomic fields. More generally Dedekind derived a formula for the class number of arbitrary number fields. This formula contains 'analytic objects', such as ζ- and L-functions, already introduced by Dirichlet in the thirties.

Dirichlet introduced functions of the form

$$L(s,\chi)=\sum_{n=1}^{\infty}\frac{\chi(n)}{n^s},$$

with complex argument s such that $\Re(s)\gg 0$ to assure convergence, to prove the famous **Theorem on Arithmetic Progressions**:

Theorem 1.2.1 (Dirichlet(1837)) *Every sequence a, $a+b$, $a+2b$, $a+3b,\ldots,$ with a and b relatively prime integers, contains an infinite number of primes.*

The $\chi(n)$ are complex numbers of absolute value one, defined on the (classes of the) integers modulo a fixed positive integer m, called the **modulus**, and with the property that $\chi(k.l)=\chi(k).\chi(l)$ for integers k and l, prime to m. By definition $\chi(n)=0$ whenever $\gcd(n,m)\neq 1$. These χ are called **Dirichlet characters** mod m. The trivial character $\chi=1$ is the one with $\chi(n)=1$ for all $n\in\mathbf{Z}$. For such a character χ mod m, the **conductor** $f=f(\chi)$ is defined as the smallest divisor of m such that there exists a character χ' mod f with the property that $\chi(n)=\chi'(n)$ for all n with $\gcd(n,m)=1$.

A fundamental property of such L-functions is their **Euler product representation**, i.e. they can be written as an infinite product over the prime numbers:

$$L(s,\chi)=\prod_p\frac{1}{\left(1-\frac{\chi(p)}{p^s}\right)},\quad \Re(s)\gg 0. \tag{1.1}$$

For s-values where the infinite sum in the definition of $L(s,\chi)$ has no meaning one uses analytic continuation, and one can prove that for a non-trivial character χ, $L(s,\chi)$ becomes an **entire** function.

A useful feature exhibited by these L-functions is their **functional equation**: Let $L(s,\chi)$ be a Dirichlet L-function with character χ mod f, where f is the conductor of χ (such χ is called **primitive**) and define

$$\Lambda(s,\chi)=\left(\frac{f}{\pi}\right)^{\frac{s}{2}}\Gamma\left(\frac{s+a}{2}\right)L(s,\chi), \tag{1.2}$$

where $a = 0$ if $\chi(-1) = 1$ and $a = 1$ if $\chi(-1) = -1$, and $\Gamma(z)$ denotes the well-known Γ-function. Then the functional equation is

$$\Lambda(s,\chi) = W(\chi)\,\Lambda(1-s,\bar{\chi}), \tag{1.3}$$

with $W(\chi) = \frac{(-i)^a}{\sqrt{f}} \sum_{n \bmod (f)} \chi(n) \exp\left(\frac{2\pi i n}{f}\right)$ is a complex number of absolute value 1, called the **root number.**

The most famous L-function arises when one takes the trivial character $\chi(n) = 1$ for all $n = 1, 2, 3, \ldots$: **Riemann's ζ-function**

$$L(s,1) = \zeta(s) = \sum_{n=1}^{\infty} \frac{1}{n^s} = \prod_p \frac{1}{(1-p^{-s})}, \ \Re(s) > 1. \tag{1.4}$$

This function plays a very important role in number theory and has kept mathematicians busy ever since its introduction by Euler, and more particularly by Riemann in 1859.

The existence of the above Euler product for $\zeta(s)$ (already known to Euler) can be interpreted as the unique factorization property of rational integers into products of primes, and this implies, as we know, that $h(\mathbf{Q}) = 1$.

After analytic continuation $\zeta(s)$ becomes a meromorphic function of s with a simple pole at $s = 1$ and residue equal to 1. The function $\zeta(s)$ has simple zeroes at the integers $s = -2m$, $m = 1, 2, 3, \ldots$ (the trivial zeroes) and infinitely many zeroes on the line $\Re(s) = \frac{1}{2}$. There is the famous conjecture:

Conjecture 1.2.1 (Riemann's Hypothesis) *All non-trivial zeroes of Riemann's ζ-function $\zeta(s)$ lie on the line $\Re(s) = \frac{1}{2}$.*

Until this day this has been neither proved nor disproved!

For an arbitrary number field K one defines the **Dedekind ζ-function** as follows:

Definition 1.2.1 *The Dedekind ζ-function of K is:*

$$\zeta_K(s) = \sum_I \frac{1}{(NI)^s}, \ \Re(s) > 1,$$

where I runs over all ideals of the ring of integers $\mathcal{O}_K$ of K, and where NI denotes the **norm** of I, i.e. the cardinality of the quotient ring $\mathcal{O}_K/I$. For $\Re(s) > 1$ one has $\zeta_K(s) \neq 0$. For $K = \mathbf{Q}$ one recovers Riemann's ζ-function. $\zeta_K(s)$ has an Euler product expansion of the form

$$\zeta_K(s) = \sum_I \frac{1}{(NI)^s} = \prod_{\wp} \frac{1}{(1-(N\wp)^{-s})}, \tag{1.5}$$

where the product is taken over all prime ideals $\wp$ of $\mathcal{O}_K$.
$\zeta_K(s)$ also has a **functional equation**. To state it, let

$$A(K) = 2^{-r_2}\,\pi^{-\frac{n}{2}}\,|D(K)|^{\frac{1}{2}},$$

where r_2 is the number of pairs of complex conjugate embeddings of K, n is the degree of K over $\mathbf{Q}$ and $D(K)$ is the **discriminant** of K, i.e. a rational integer equal to the determinant of the traces, $\det(\mathrm{Tr}_{\mathcal{O}_K/\mathbf{Z}}(x_i x_j))$, where $x_1, \ldots, x_n$ is a basis of the $\mathbf{Z}$-module $\mathcal{O}_K$. To fix ideas, one may think of $D(K)$ as the dicriminant (up to a factor) of the minimum polynomial defining the field K. For the quadratic fields $\mathbf{Q}(\sqrt{d})$, d a square free integer, one has:

$$D(\mathbf{Q}(\sqrt{d})) = \begin{cases} d, & \text{if } d \equiv 1 \pmod 4 \\ 4d, & \text{if } d \equiv 2,3 \pmod 4 \end{cases} \tag{1.6}$$

and for the cyclotomic fields $\mathbf{Q}(\zeta_p)$, p an odd prime number, the discriminant is given by:

$$D(\mathbf{Q}(\zeta_p)) = (-1)^{\frac{p-1}{2}}\,p^{p-2}. \tag{1.7}$$

Define $Z_K(s) = A(K)^s\,\Gamma\left(\frac{s}{2}\right)^{r_1}\,\Gamma(s)^{r_2}\,\zeta_K(s)$. Here r_1 is the number of real embeddings of K. Then one has the functional equation

$$Z_K(s) = Z_K(1-s). \tag{1.8}$$

1.3 The Class Number Formula

To state the **Class Number Formula** we need a few more notions related to the units in a number field K. First, a **unit** is just an invertible element in the ring of integers $\mathcal{O}_K$ of K. The units form a multiplicative group $\mathcal{O}_K^*$ in $\mathcal{O}_K$. Special units are the roots of unity contained in K, but there may be others. Of fundamental importance is

Theorem 1.3.1 (Dirichlet's Unit Theorem) *The group of units of $\mathcal{O}_K$ is isomorphic to the direct sum of the group of roots of unity in K and a free part, isomorphic to $\mathbf{Z}^{r_1+r_2-1}$, where r_1 and r_2 are the numbers of real and pairs of complex conjugate embeddings of K, respectively.*

Thus the **degree** of K (over $\mathbf{Q}$) is $r_1 + 2r_2$. Generators of the free part are called **fundamental units**.

Next, there is the notion of the **regulator** of K. To define it, take a basis of fundamental units $\varepsilon_1, \varepsilon_2, \ldots, \varepsilon_r$, with $r = r_1 + r_2 - 1$, and consider the vectors

$$v_i = (\log|\sigma_1(\varepsilon_i)|, \ldots, \log|\sigma_{r_1}(\varepsilon_i)|, \log|\sigma_{r_1+1}(\varepsilon_i)|^2, \ldots, \log|\sigma_{r_1+r_2}(\varepsilon_i)|^2)$$

in $\mathbf{R}^{r_1+r_2}$, $i = 1, 2, \ldots, r$. The $\sigma_1, \ldots, \sigma_{r_1}$ denote the real embeddings, and the $\sigma_{r_1+1}, \ldots, \sigma_{r_1+r_2}$ are the complex embeddings (up to conjugates) of K. By the so-called **product formula** one sees that the v_i are all contained in the hyperplane $x_1 + x_2 + \cdots + x_{r_1+r_2} = 0$ in $\mathbf{R}^{r_1+r_2}$ (with respect to the standard basis). The **regulator** R is now defined to be $\dfrac{1}{r_1+r_2}$ times the volume of the parallelepiped, spanned by the v_i's and the vector $(1, 1, \ldots, 1) \in \mathbf{R}^{r_1+r_2}$. One can show that

$$R = |\det(a.\log|\sigma_i(\varepsilon_j)|)|_{1\le i,j\le r} \tag{1.9}$$

for any choice of r embeddings σ out of the $r+1$. Besides $a = 1$ for real σ and $a = 2$ for complex σ. Thus R is a positive **real** number, and, by definition, $R = 1$ if $r = 0$. Finally, let w denote the number of roots of unity in K and $D = D(K)$ is the discriminant of K.
The relation between all these numbers is now given by **Dedekind's Class Number Formula**

Theorem 1.3.2 (Dedekind) *After analytic continuation, $\zeta_K(s)$ becomes a meromorphic function with a simple pole at $s = 1$ and residue*

$$\lim_{s\to 1}(s-1)\,\zeta_K(s) = \frac{2^{r_1}(2\pi)^{r_2}R}{w\sqrt{|D|}} \cdot h(K)\ .$$

Example 1.3.1 For a real quadratic field $\mathbf{Q}(\sqrt{d})$, $d > 0$ square free, one has $r_1 = 2$, $r_2 = 0$, $R = \log \varepsilon$, where ε is a fundamental unit, $w = 2$ and finally:

$$\lim_{s \to 1}(s-1)\,\zeta_K(s) = \frac{2\log\varepsilon}{\sqrt{D}} \cdot h(\mathbf{Q}(\sqrt{d})).$$

Example 1.3.2 For an imaginary quadratic field $\mathbf{Q}(\sqrt{d})$, $d < 0$ square free, one has $r_1 = 0$, $r_2 = 1$, $R = 1$, $w = 4$ if $d = -1$, $w = 6$ if $d = -3$ and $w = 2$ for other values of d. The result is:

$$\lim_{s \to 1}(s-1)\,\zeta_K(s) = \frac{2\pi}{w\sqrt{|D|}} \cdot h(\mathbf{Q}(\sqrt{d})).$$

Example 1.3.3 For the cyclotomic field $\mathbf{Q}(\zeta_p)$, p an odd prime, one finds:

$$\lim_{s \to 1}(s-1)\,\zeta_K(s) = \frac{2^{\frac{p-3}{2}}\,\pi^{\frac{p-1}{2}}\,R}{p^{p/2}} \cdot h(\mathbf{Q}(\zeta_p)).$$

Units and R were calculated by Kummer for many p's. In general it is rather difficult to obtain all units of a cyclotomic field, but a subgroup of finite index of the group of all units in $\mathbf{Q}(\zeta_p)$ is given by the so-called **cyclotomic units**. This group of cyclotomic units is generated by -1, ζ_p and $\pm\dfrac{\sin(\pi a/p)}{\sin(\pi/p)}$, with $1 < a < \dfrac{p}{2}$, $\gcd(a,p) = 1$. For interesting results on cyclotomic units and their importance one of the best references is [Wal].

Remark 1.3.1 Using the functional equation one can simplify the class number formula to get:

$$\lim_{s \to 0} \zeta_K(s)\, s^{-(r_1+r_2-1)} = -\frac{R}{w} \cdot h(K) \tag{1.10}$$

This formula will play an important role when we will discuss it in a more general context to study values of L-functions at integer values of the argument s in terms of K-theory. The relevant facts will be that the order of $\zeta_K(s)$ at $s = 0$ is equal to the **rank** of the **group of units** (or the $\mathbf{Q}$-dimension after tensoring with $\mathbf{Q}$) and the first non-zero coefficient of the Taylor series expansion of $\zeta_K(s)$ at $s = 0$ equals the value of the **regulator** up to a non-zero rational number.

1.4 Abelian Number Fields

For the actual calculation of the class number one would like to have a concrete means of handling the left hand side of Dedekind's class number formula. This is provided by L-functions. In the sequel we assume that $K/\mathbf{Q}$ is a **normal extension**, so there is a well defined **Galois group** $G = \mathrm{Gal}(K/\mathbf{Q})$. $\mathrm{Gal}(K/\mathbf{Q})$ is the group of automorphisms of K. They leave each element of $\mathbf{Q}$ fixed. For a finite Galois extension $K/\mathbf{Q}$ the order of $\mathrm{Gal}(K/\mathbf{Q})$ is equal to the degree of the number field $K/\mathbf{Q}$.

To see how (and what kind of) L-functions enter the scene, we distinguish between two cases: G abelian and G non-abelian. K is called an **abelian number field** if $G = \mathrm{Gal}(K/\mathbf{Q})$ is abelian, otherwise it is called **non-abelian**. In this section we consider the abelian case.

So let K be an abelian number field of degree n. Then $\mathrm{Gal}(K/\mathbf{Q})$ admits n different characters $\chi_i : G \longrightarrow \mathbf{C}^*$, $i = 1, 2, \ldots, n$, taking values in the roots of unity. One of these characters is the trivial one, taking the value $+1$ for all $\sigma \in G$.

A first example is furnished by the cyclotomic field $\mathbf{Q}(\zeta_p)$, where p is an odd prime. Here the Galois group is isomorphic with the multiplicative group $(\mathbf{Z}/p\mathbf{Z})^*$. Under this isomorphism the characters of $\mathrm{Gal}(\mathbf{Q}(\zeta_p)/\mathbf{Q})$ are the trivial one and $p-2$ non-trivial Dirichlet characters mod p. More generally, let $\mathbf{Q}(\zeta_m)$, $m \geq 3$, denote the mth cyclotomic field, obtained by adjoining a primitive m^{th} root of unity ζ_m to $\mathbf{Q}$. Then $\mathbf{Q}(\zeta_m)$ has degree $\phi(m)$, where ϕ is Euler's totient function, and $\mathrm{Gal}(\mathbf{Q}(\zeta_m)/\mathbf{Q})$ is isomorphic with the multiplicative abelian group of reduced residue classes mod m, $(\mathbf{Z}/m\mathbf{Z})^*$. Under this isomorphism there are $\phi(m) - 1$ non-trivial Dirichlet characters mod m. For their determination we refer to [Co].

For real and imaginary quadratic number fields $\mathbf{Q}(\sqrt{d})$, $d \in \mathbf{Z}$ square free, the Galois group consists of two elements: the identity and the conjugation map which sends an element $a + b\sqrt{d}$ to $a - b\sqrt{d}$, $a, b \in \mathbf{Q}$. The non-trivial character will be of order two, in other words, it takes the values ± 1. Using Dirichlet's Lemma (cf. [Co]), Remark 1.4.1 and Theorem 1.4.2 below, one deduces that the corresponding χ of Theorem 1.4.2 is the Dirichlet character χ mod $|D|$, where D is the discriminant, given by the following expressions:

$$\chi(k) = \left(\frac{D}{k}\right) = \begin{cases} \left(\frac{k}{|D|}\right), & \text{if } d \equiv 1 \pmod 4 \\ (-1)^{\frac{k-1}{2}} \left(\frac{k}{|D|}\right), & \text{if } d \equiv 3 \pmod 4 \\ (-1)^{\frac{k^2-1}{8} \cdot \frac{k-1}{2} \cdot \frac{d'-1}{2}} \left(\frac{k}{|d'|}\right), & \text{if } d = 2d' \ (d' \text{odd}) \end{cases} \tag{1.11}$$

where $\gcd(D, k) = 1$, otherwise $\chi(k) := 0$, and where $\left(\frac{k}{\cdot}\right)$ is the Jacobi symbol.

At this stage we invoke the Kronecker-Weber Theorem from class field theory (cf. [Ne]).

Theorem 1.4.1 (Kronecker-Weber) *Every finite abelian extension K of $\mathbf{Q}$ is contained in a cyclotomic extension $\mathbf{Q}(\zeta_m)$.*

Remark 1.4.1 m need not be a prime. The smallest m in the theorem is called the conductor of K. For quadratic fields one can prove that the conductor is equal to the conductor of the non-trivial character, which equals the absolute value of the discriminant D of the field.

Remark 1.4.2 The Dirichlet character $\left(\frac{D}{\cdot}\right)$ is a character of $\mathbf{Q}(\zeta_{|D|})$.

As a corollary of the theorem one obtains an exact sequence of abelian groups:

$$1 \longrightarrow \mathrm{Gal}(\mathbf{Q}(\zeta_m)/K) \longrightarrow \mathrm{Gal}(\mathbf{Q}(\zeta_m)/\mathbf{Q}) \longrightarrow \mathrm{Gal}(K/\mathbf{Q}) \longrightarrow 1, \tag{1.12}$$

and one sees that a character of $\mathrm{Gal}(\mathbf{Q}(\zeta_m)/\mathbf{Q})$ induces one on the relative Galois group $H = \mathrm{Gal}(\mathbf{Q}(\zeta_m)/K)$ of automorphisms of $\mathbf{Q}(\zeta_m)$ leaving the elements of K fixed.

We can now state the fundamental theorem relating the ζ-function $\zeta_K(s)$ to L-functions.

Theorem 1.4.2 *Let K be a finite abelian number field with conductor m. Then:*

$$\zeta_K(s) = \prod_{\chi|H=1} L(s,\chi),$$

where the product is taken over all characters χ of $\mathrm{Gal}(\mathbf{Q}(\zeta_m)/\mathbf{Q})$ *which induce the trivial one on* $H = \mathrm{Gal}(\mathbf{Q}(\zeta_m)/K)$.

We know that $L(s,1) = \zeta(s)$, Riemann's ζ-function, has a simple pole at $s = 1$ with residue 1, and that the other $L(s,\chi)$ are entire functions. So we may write

$$\zeta_K(s) = \zeta(s) \prod_{\chi|H=1}{}' L(s,\chi), \tag{1.13}$$

where $'$ means that the trivial character is to be excluded. For the class number formula one gets the result:

Theorem 1.4.3 *Let K be a finite abelian number field. Then:*

$$h(K) = \frac{w\sqrt{|D|}}{2^{r_1}(2\pi)^{r_2} R} \cdot \prod_{\chi|H=1}{}' L(1,\chi).$$

Special cases are the quadratic and the cyclotomic fields. For the real quadratic field $\mathbf{Q}(\sqrt{d})$, $d > 0$ square free, one has

$$h(\mathbf{Q}(\sqrt{d})) = \frac{\sqrt{D}}{2\log\varepsilon} \cdot L(1, \left(\frac{D}{\cdot}\right)) \tag{1.14}$$

For the imaginary quadratic field $\mathbf{Q}(\sqrt{d})$, $d < 0$ square free, one finds

$$h(\mathbf{Q}(\sqrt{d})) = \frac{w\sqrt{|D|}}{2\pi} \cdot L(1, \left(\frac{D}{\cdot}\right)) \tag{1.15}$$

Actually, these formulae can be much further reduced to give

$$h(\mathbf{Q}(\sqrt{d}) = \frac{-1}{\log\varepsilon} \sum_{0<x<D/2} \left(\frac{D}{x}\right) \log\sin\left(\frac{\pi x}{D}\right), \; d > 0, \tag{1.16}$$

and

$$h(\mathbf{Q}(\sqrt{d}) = \frac{w}{2D}\sum_{x=1}^{-D}\left(\frac{D}{x}\right) = \frac{1}{2-\left(\frac{D}{2}\right)}\sum_{0<x<-D/2}\left(\frac{D}{x}\right), \; d<0. \quad (1.17)$$

(For this last equality one must assume that $D < -4$.)
For the cyclotomic field $\mathbf{Q}(\zeta_p)$, p an odd prime, one gets

$$h(\mathbf{Q}(\zeta_p)) = \frac{w\,p^{\frac{p-2}{2}}}{(2\pi)^{\frac{p-1}{2}}R}\cdot\prod{}' L(1,\chi) \quad (1.18)$$

An intensive study of the class number formulae for abelian number fields can be found in [Ha].

1.5 Non-abelian Number Fields and Artin L-Functions

For non-abelian number fields things are more complicated. There is no such a simple description of L-functions by means of characters taking their values in the roots of unity. Instead, one introduces so-called Artin L-functions, associated to (irreducible) representations of the non-abelian Galois group $G = \mathrm{Gal}(K/\mathbf{Q})$. The abelian case will correspond to one-dimensional representations, i.e. characters.

So, let K be a number field (finite and Galois as always) with ring of integers $\mathcal{O}_K$ and Galois group $G = \mathrm{Gal}(K/\mathbf{Q})$. Let $\rho : G \longrightarrow \mathrm{Aut}(V) = GL(V)$ be a finite dimensional complex representation. Take a prime ideal $\wp$ in $\mathcal{O}_K$ lying over the prime number $p \in \mathbf{Z}$. For such a $\wp$ a subgroup $D_\wp$ is defined as follows:

Definition 1.5.1 *The* decomposition group $D_\wp$ *of* $\wp$ *over* p *is*

$$D_\wp = \{\sigma \in G | \sigma(\wp) = \wp\}.$$

Now an element $\sigma \in D_\wp$ defines a $\bar{\sigma} : \mathcal{O}_K/\wp \longrightarrow \mathcal{O}_K/\wp$, i.e. an element of the Galois group of the residue field $\mathcal{O}_K/\wp \cong \mathbf{F}_q$, $q = p^f$, over the finite field $\mathbf{Z}/p\mathbf{Z} \cong \mathbf{F}_p$, but this latter group is generated by the Frobenius element $Fr_p : x \mapsto x^p$.

Definition 1.5.2 *The* inertia group $I_\wp \lhd D_\wp$ *of* $\wp$ *is the normal subgroup of* $D_\wp$ *consisting of all* $\sigma \in D_\wp$ *such that* $\bar{\sigma} = id.$.

Thus one obtains an exact sequence

$$1 \longrightarrow I_\wp \longrightarrow D_\wp \longrightarrow \langle Fr_p \rangle \longrightarrow 1, \tag{1.19}$$

in other words, $\langle Fr_p \rangle \cong D_\wp / I_\wp$. This implies that the representation ρ induces an action of Fr_p on $V^{I_\wp}$, the $I_\wp$-invariants of V. We write $\rho(Fr_p)$ for this action. Fr_p is called the arithmetic Frobenius automorphism , while its inverse, which plays an even more important role in the study of L-functions, is called the geometric Frobenius. We will denote it by F_p. It should be observed that this whole construction does not depend on the choice of $\wp$ over p. Of course it depends on the prime number p.

Definition 1.5.3 *The* p^{th} *factor* $L_p(s,\rho)$ *of* Artin's L-function

$$L(s,\rho) = \prod_p L_p(s,\rho)$$

is defined as follows:

$$L_p(s,\rho) = \frac{1}{\det\left(1 - \rho(Fr_p).p^{-s}|V^{I_\wp}\right)}, \quad \Re(s) > 1.$$

(ρ, V) *is called* unramified at p *if* $V^{I_\wp} = V$, *i.e.* $\rho|I_\wp = id.$, *otherwise it is called* ramified at p .

The generalization of Theorem 1.4.2 is given by

Theorem 1.5.1 (Artin) *For an arbitrary (finite) number field* K *one has the product expansion*

$$\zeta_K(s) = \prod_{i=1}^{N} L(s,\rho_i)^{\dim V_i},$$

where the (ρ_i, V_i) *run over all (irreducible) representations of the Galois group* $\mathrm{Gal}(K/\mathbf{Q})$.

The compatibility of Theorem 1.4.2 and Theorem 1.5.1 follows from a special case of Artin's Reciprocity Theorem, part of which is

Theorem 1.5.2 (Artin) *For an* abelian *number field $K/\mathbf{Q}$ with one-dimensional representation ρ of* $\mathrm{Gal}(K/\mathbf{Q})$ *given by multiplication by the Dirichlet character χ, Artin's L-function $L(s, \rho)$ coincides with the Dirichlet L-function $L(s, \chi)$.*

The character of the representation ρ is the function $\chi : G \longrightarrow \mathbf{C}$, with $\chi(\sigma) = \operatorname{Tr} \rho(\sigma)$. Then χ is a central function, which means that χ depends only on the equivalence class of the representation ρ, or, in other words, χ characterizes the representation up to isomorphism. Therefore one often writes $L(s, \chi)$ instead of $L(s, \rho)$. For the trivial character $\chi = 1$, in this latter sense, one recovers Riemann's ζ-function: $L(s, 1) = \zeta(s)$. Furthermore, for two characters χ_1 and χ_2 one has

$$L(s, \chi_1 + \chi_2) = L(s, \chi_1)\, L(s, \chi_2) \tag{1.20}$$

As to the analytic behaviour of Artin's L-functions, the situation is far less understood than in the case of Dirichlet L-functions. More precisely, there is a long standing conjecture:

Conjecture 1.5.1 (Artin's Conjecture) *For every irreducible representation ρ with non-trivial character χ, Artin's L-function $L(s, \chi) = L(s, \rho)$ extends to an* entire function *on the complex plane.*

R. Brauer proved in 1947 that $L(s, \chi)$ extends to a meromorphic function on the complex plane. E. Artin (1898–1962) was able to prove his conjecture for representations induced from one-dimensional representations of a subgroup of G. These are called the monomial representations. Actually, Artin introduced his L-functions and stated his conjecture with a non-abelian generalization of class field theory in mind and his proof of the monomial case can be considered as a version of the Reciprocity Law of abelian class field theory.

The proof for higher dimensional representations is part of the so-called Langlands Program, where the philosophy is that every L-function is a product of automorphic L-functions. As a matter of fact, R. Langlands proved Artin's Conjecture for a whole class of 2-dimensional representations of the Galois group. By a result of Serre [Se2] one knows that the image of a 2-dimensional representation ρ of $G = \mathrm{Gal}(K/\mathbf{Q})$ in $PGL(2, \mathbf{C})$ is either

(*i*) D_n, the dihedral group, in which case ρ is monomial;
(*ii*) A_4, the tetrahedral group;
(*iii*) S_4, the octahedral group;
(*iv*) A_5, the icosahedral group.

Langlands was able to prove Artin's Conjecture in the tetrahedral case and, under certain conditions, in the octahedral case. However, his methods (automorphic forms on $GL(2)$ and $GL(3)$) do not apply to the icosahedral case, which remains out of reach at this moment, (cf. [Ge2]).

Artin L-functions admit a functional equation. To state it we need a few more notions. First, there are the higher ramification groups $G_i \subset D_\wp$, $i = 0, 1, 2, \ldots$. G_i is defined as the normal subgroup which act trivially on $\mathcal{O}_K/\wp^{i+1}$. In particular, $G_0 = I_\wp$, the inertia group.

Definition 1.5.4 *The* Artin conducter *is defined as*

$$f = f(\rho) = f(\chi) = \prod_p p^{n_p(\chi)},$$

where $n_p(\chi) = \sum_i \frac{g_i}{g_0} \operatorname{codim}(V)^{G_i}$, *and* $g_i = \#G_i$.

Theorem 1.5.3 *The* $n_p(\chi)$*'s are* integers.

The functional equation takes the form

$$\Lambda(s, \chi) = W(\chi)\,\Lambda(1 - s, \bar{\chi}), \tag{1.21}$$

where

$$\Lambda(s, \chi) = (f(\chi)\pi^{-\dim(V)})^{\frac{s}{2}}\, \Gamma\left(\frac{s+1}{2}\right)^a \Gamma\left(\frac{s}{2}\right)^{\dim(V)-a}, \tag{1.22}$$

with $a = \operatorname{codim}(V^{G_\infty})$. Here we have written G_∞ for the group generated by complex conjugation at an infinite prime of K, thus G_∞ has order 1 (resp. 2) at a real (resp. complex) place of K. Actually, a is independent of the choice of the real (resp. complex) place ∞ lying over the real place ∞ of $\mathbf{Q}$. $\bar{\chi}$ is the character corresponding to the contragredient representation $\bar{\rho} : G \longrightarrow GL(V^*)$, where V^* is the dual of V. $\bar{\rho}$ is defined by $< \bar{\rho}(\sigma)f, v >=< f, \rho(\sigma)^{-1}v >$, $v \in V$, $f \in V^*$ and $\sigma \in G$, (cf. [Ma]). $W(\chi)$ is a constant of absolute value 1 and is called **Artin's root number**. For two characters χ_1 and χ_2 the root numbers satisfy

$$W(\chi_1 + \chi_2) = W(\chi_1)\, W(\chi_2) \tag{1.23}$$

Example 1.5.1 For quadratic number fields $\mathbf{Q}(\sqrt{d})$, d a square free integer, χ a Dirichlet character of conductor f, it is not hard to prove that $\chi(-1) = 1$ (resp. -1) if $d > 0$ (resp. $d < 0$) and using equation (1.11) one regains formula (1.3).

To finish this chapter we give three properties of Artin L-functions:

1. $\log L_p(s, \rho) = \sum_{n=1}^{\infty} \frac{\mathrm{Tr}(\rho(Fr_p)^n | V^{I_p})}{n.p^{ns}}$;
2. $L_p(s, \rho)$ is analytic for $\Re(s) > 0$;
3. $L_p(s, \rho) \neq 0$ whenever $\Re(s) > 1$.

Remark 1.5.1 With **Beilinson's Conjectures** on the values of L–functions of arbitrary varieties (or **motives**) over $\mathbf{Q}$ in mind, we will need a generalization of the notion of an Artin L-function. To this end, we note that the set-up of Artin's L-functions extends to **infinite** Galois extensions, in particular the absolute extension $\bar{\mathbf{Q}}/\mathbf{Q}$. One introduces a suitable topology on the Galois group

$$G = \mathrm{Gal}(\bar{\mathbf{Q}}/\mathbf{Q}) = \varprojlim \mathrm{Gal}(K/\mathbf{Q}), \tag{1.24}$$

where K runs over the finite Galois extensions of $\mathbf{Q}$. Then G is compact and totally disconnected in this so-called **Krull topology**. By a representation we will mean a **continuous** finite dimensional representation of G in the ℓ-adic Lie group $\mathrm{Aut}(V)$, where V is a finite dimensional vector space over the ℓ-adic numbers $\mathbf{Q}_\ell$, with induced action of the geometric Frobenius.

Chapter 2

The one-dimensional case: elliptic curves

The next class of objects we consider concerns one-dimensional varieties over the rational numbers, i.e. curves over $\mathbf{Q}$. *The simplest examples are the projective line* $\mathbf{P}^1$ *and conics over* $\mathbf{Q}$. *From the point of view of algebraic geometry these are equivalent objects and they are well understood. For our purposes we note that they are curves of genus zero and contain an infinite number of rational points (if any). More interesting are the elliptic curves over* $\mathbf{Q}$. *These have been the subject of deep study from various points of view, algebraic, arithmetic and geometric. From the arithmetic point of view they give rise to some of the most intricate conjectures, the Birch & Swinnerton-Dyer Conjectures, which can be interpreted as the one-dimensional counterpart of Dedekind's Class Number Formula. Also, more recently, a remarkable relation was found between elliptic curves and Fermat's Last Theorem.*

2.1 General Features of Elliptic Curves

By an **elliptic curve** E defined over the rationals $\mathbf{Q}$ we mean a smooth projective curve of genus one, equipped with a $\mathbf{Q}$-rational point O, the **origin**. Such a curve admits a plane projective model in $\mathbf{P}^2(\bar{\mathbf{Q}})$, given by the equation

$$E : Y^2Z + a_1XYZ + a_3YZ^2 = X^3 + a_2X^2Z + a_4XZ^2 + a_6Z^3 . \quad (2.1)$$

The origin O corresponds to the point $(0:1:0) \in \mathbf{P}^2(\bar{\mathbf{Q}})$. Restricting to the affine part, one gets the so-called (generalized) Weierstraß equation:

$$E: y^2 + a_1xy + a_3y = x^3 + a_2x^2 + a_4x + a_6, \tag{2.2}$$

with $a_1, \ldots, a_6 \in \mathbf{Q}$. This can be further transformed into

$$E: y^2 = x^3 + Ax + B, \text{ or } E: y^2 = 4x^3 - g_2x - g_3, \tag{2.3}$$

for suitable A, B, g_2 and g_3.

Elliptic curves are particularly interesting objects, because apart from their beautiful geometric properties they also have special algebraic and arithmetic features which distinguish them from all other curves. They constitute the one-dimensional examples of so-called abelian varieties, i.e. (non-singular) projective algebraic varieties with an abelian group structure on the set of their points. The origin plays the role of the zero-element with respect to this group structure on the elliptic curve E. Restricting to the real points of the curve, one can visualize very easily the group structure on E, e.g. three points of E lie on a line if and only if their sum is equal to the zero-element O.

For two elliptic curves E and E' with origins O and O' defined over $\mathbf{Q}$, and an extension K of $\mathbf{Q}$ which is contained in $\bar{\mathbf{Q}}$, one defines a K-homomorphism of E to E' as a rational map from E to E', defined over K, that is a group homomorphism. In fact, any rational map from E to E' which is defined over K and which sends the origin O to O' is a K-homomorphism. Similarly one has the notion of K-isomorphism between two elliptic curves and of K-automorphism of one such curve.

Given an equation of the form (2.2), when does it define an elliptic curve? To answer this question one introduces the so-called discriminant $\Delta = \Delta(E)$ of E as follows. Write

$$b_2 = a_1^2 + 4a_2, \quad b_4 = a_1a_3 + 2a_4, \quad b_6 = a_3^2 + 4a_6,$$

$$b_8 = \frac{b_2b_6 - b_4^2}{4} = a_1^2a_6 + 4a_2a_6 - a_1a_3a_4 + a_2a_3^2 - a_4^2,$$

$$c_4 = b_2^2 - 24b_4, \quad c_6 = -b_2^3 + 36b_2b_4 - 216b_6.$$

Then the discriminant $\Delta(E)$ is given by the expression

$$\Delta(E) = \frac{c_4^3 - c_6^2}{12^3} = -b_2^2b_8 - 8b_4^3 - 27b_6^2 + 9b_2b_4b_6. \tag{2.4}$$

In terms of A and B (or g_2 and g_3) the discriminant becomes

$$\Delta(E) = -16(4A^3 + 27B^2) \quad (\text{or } \Delta(E) = g_2^3 - 27g_3^2)\,. \tag{2.5}$$

Then Weierstraß's equation defines an elliptic curve over $\mathbf{Q}$ if and only if $\Delta(E) \neq 0$. Actually this is equivalent to the non-singularity of E. Also, let the j-**invariant** of E, $j = j(E)$, be defined by

$$j(E) = \frac{c_4^3}{\Delta(E)}\,. \tag{2.6}$$

In terms of A and B (or g_2 and g_3) one can write

$$j(E) = \frac{12^3(4A)^3}{\Delta(E)} \quad (\text{or } j(E) = 12^3\frac{g_2^3}{\Delta(E)})\,. \tag{2.7}$$

The j-invariant sets up a bijection between the set of $\bar{\mathbf{Q}}$–isomorphism classes of elliptic curves over $\mathbf{Q}$ and the one-dimensional affine space $\mathbf{A}^1(\mathbf{Q})$ over $\mathbf{Q}$. Furthermore, for given $j \in \mathbf{Q}$, the set of $\mathbf{Q}$–isomorphism classes of elliptic curves over $\mathbf{Q}$ with j-invariant j is equal to the cohomology set $H^1(\mathrm{Gal}(\bar{\mathbf{Q}}/\mathbf{Q}), \mathrm{Aut}(E))$, where $\mathrm{Aut}(E)$ is the group of $\bar{\mathbf{Q}}$–automorphisms of any elliptic curve E over $\mathbf{Q}$ with j-invariant j.

Because we are interested in arithmetic properties of elliptic curves defined over $\mathbf{Q}$ we shall need to consider versions of such curves over local fields and over finite fields. In particular the endomorphism ring of the curve (to be defined below) will play an important role on several places. So in this paragraph we consider elliptic curves E and E' over an arbitrary field k. The definition of an elliptic curve over k is similar to the one given for elliptic curves over $\mathbf{Q}$, and all related notions introduced in this section admit a straightforward generalization to the case of elliptic curves over arbitrary fields k. The simplified Weierstraß equation (2.3) is valid for k with $\mathrm{char}(k) \neq 2, 3$.

Let E (with origin O) and E' (with origin O') be two elliptic curves defined over the field k. A non-zero $\bar{k}$–homomorphism $\lambda : E \longrightarrow E'$ is called an **isogeny** between E and E'. A theorem in algebraic geometry says that λ is surjective. The isogeny λ gives rise to an embedding of the function field $\bar{k}(E')$ into the function field $\bar{k}(E)$, and one defines the **degree** of λ, $\deg(\lambda)$, as follows:

$$\deg(\lambda) = [\bar{k}(E) : \bar{k}(E')] = [\bar{k}(E) : \bar{k}(E')]_s . [\bar{k}(E) : \bar{k}(E')]_i \tag{2.8}$$

where $\deg(\lambda)_s$ is the **separable degree** of λ, which turns out to be number of points in $\lambda^{-1}(Q)$, for every $Q \in E'$, and where $\deg(\lambda)_i$ is the purely inseparable degree of λ, also called the **ramification index** of λ. Actually, an isogeny $\lambda : E \longrightarrow E'$ can be shown to be a group homomorphism with finite kernel $\lambda^{-1}(O')$ of order $\deg(\lambda)_s$. For convenience one also allows the **trivial isogeny** $o : E \longrightarrow E'$, which by definition maps E to $O' \in E'$. Its degree is defined to be zero.

Given two isogenies $\lambda, \mu : E \longrightarrow E'$, their sum is defined by

$$(\lambda + \mu)(P) = \lambda(P) + \mu(P)\,, \; P \in E\,,$$

thus giving a group structure on the set of isogenies between E and E'. The zero element is given by the trivial isogeny. The resulting group is denoted by $\mathrm{Hom}(E, E')$. Similarly one can define $\mathrm{Hom}_k(E, E')$ for the group of isogenies between E and E', defined over k. In case $E = E'$ one can define a multiplication on $\mathrm{Hom}(E, E')$, given by composition. Thus one obtains a rather interesting invariant of the curve, the **endomorphism ring** $\mathrm{End}(E)$. Similarly, one defines $\mathrm{End}_k(E)$. The invertible elements in this ring form the group of **automorphisms** of E, $\mathrm{Aut}(E)$ and similarly $\mathrm{Aut}_k(E)$.

An important example of an isogeny in $\mathrm{End}(E)$ is **multiplication** by m, $m \in \mathbf{Z}$, defined as $[m] : E \longrightarrow E$, with $[m](P) = P + P + \cdots + P$ (m times) if $m > 0$, and $[m](P) = [-m](-P)$ if $m < 0$. By definition, $[0](P) = O$, the origin of E. One can prove that the isogeny $[m]$ is non-constant. This means that $\mathrm{End}(E)$ contains the ring $\mathbf{Z}$.

Given two elliptic curves E andE' (over the field k) and a non-constant isogeny $\lambda : E \longrightarrow E'$ of degree m, there exists a unique **dual isogeny**

$$\check{\lambda} : E' \longrightarrow E\,,$$

such that $\check{\lambda}\lambda = [m]$ on E, $\lambda\check{\lambda} = [m]$ on E', $\deg(\check{\lambda}) = m$ and $\check{\check{\lambda}} = \lambda$. On the ring $\mathrm{End}(E)$ the involution $\check{}$ is called the Rosati involution. For the isogeny $[m] \in \mathrm{End}(E)$ one has $[\check{m}] = [m]$ and $\deg[m] = m^2$.

For each prime number p one may consider the Weierstraß equation for E over the local field $\mathbf{Q}_p$. By a simple 'rescaling' of x and y, one can find a Weierstraß equation for E with all the coefficients in the ring of integers $\mathbf{Z}_p$. Thus at p the discriminant $\Delta_p = \Delta_p(E)$ is a p-adic integer as well, i.e. $\mathrm{ord}_p(\Delta_p) \geq 0$. So it makes sense to look for a minimal Δ_p, under the condition that the coefficients remain p-adic integers. One can show that every elliptic curve has a so-called **minimal Weierstraß equation** at each prime p, i.e. an equation such that $\text{ord}_p(\Delta_p)$ is minimal and the coefficients are p-adic integers. Actually, there is an algorithm to obtain this minimal equation, (cf. [Ta4]).

Back to the elliptic curve E over the global field $\mathbf{Q}$, the following result can be proven

Theorem 2.1.1 *Every elliptic curve E defined over $\mathbf{Q}$ has a* **global minimal Weierstraß equation** , *i.e. an equation of the form:*

$$E : y^2 + a_1xy + a_3y = x^3 + a_2x^2 + a_4x + a_6$$

with **integer coefficients** $a_1, \ldots, a_6$, *and which is* **minimal** *for all primes* p.

One defines the **minimal discriminant** $D(E)$ of $E/\mathbf{Q}$ as the discriminant of the global minimal Weierstraß equation of E. Then $D(E)$ is equal (up to sign) to the product $\prod_p p^{\text{ord}_p(\Delta_p)}$, where the product runs over the finitely many primes where the local (p-adic) discriminant $\Delta_p(E)$ has positive order.

The corresponding projective formulation of the global minimal Weierstraß equation for $E/\mathbf{Q}$ is evident.

In the sequel we will always assume that $E/\mathbf{Q}$ is given by its minimal Weierstraß equation, thus $\Delta(E) = D(E)$.

This global equation may be reduced modulo arbitrary primes. Denote the reduction mod$p$ by E_p. In this way we get, for all primes not dividing the discriminant $\Delta(E)$, an elliptic curve E_p over the finite field $\mathbf{F}_p$. For the primes that divide $\Delta(E)$ we get a singular curve with a **node** or a **cusp** . There are two kinds of nodes, the ones with rational tangents and those where the tangents are irrational over $\mathbf{F}_p$, but become rational over a quadratic extension of $\mathbf{F}_p$. One says that $E/\mathbf{Q}$ has **good reduction at** p if E_p is elliptic, otherwise $E/\mathbf{Q}$ is said to have **bad reduction at** p. In this latter case E is said to have **multiplicative** or **semi-stable reduction** if the singular point is a node. When the singular point is a cusp E is said to have **additive** or **unstable reduction** at P. Actually, J. Tate proved that any elliptic curve over $\mathbf{Q}$ has bad reduction at some prime, i.e. $\Delta(E)$ is never equal to ± 1.

2.2 Varieties over Finite Fields

We need a small intermezzo on ζ-functions of varieties over finite fields to justify the introduction of the Z-function of an elliptic curve.

In his thesis (1921) E. Artin observed that algebraic congruences in two variables modulo a prime p can be interpreted as algebraic equations over the prime field $\mathbf{F}_p$ and that there is a remarkable analogy between the polynomial ring $\mathbf{F}_p[X]$ and the ring $\mathbf{Z}$ of rational integers. To exploit this analogy, Artin studied the **integral closure** A of the ring $\mathbf{F}_p[X]$ in the splitting field of the polynomial $Y^2 - f(X) \in \mathbf{F}_p[X, Y]$, where $f(X) \in \mathbf{F}_p[X]$. This ring A is a Dedekind domain, so every ideal has a unique factorization into a product of prime ideals and every prime ideal is maximal. The set of ideals modulo principal ideals is a finite abelian group under multiplication, the so-called **ideal class group**, whose order is called, again, the **class number**. This situation is completely analogous to the number field case. This led Artin to define the ζ-function $\zeta_C(s)$ of the curve $C : Y^2 - f(X) = 0,\ f(X) \in \mathbf{F}_p[X]$, as follows:

Definition 2.2.1 *Let $C : Y^2 - f(X) = 0,\ f(X) \in \mathbf{F}_p[X]$, then the* **$\zeta$-function** *of C is*

$$\zeta_C(s) = \sum_I \frac{1}{(NI)^s}, \quad \Re(s) \gg 0,$$

where NI is the **norm** *of the ideal $I \subset A$, i.e. the cardinality of the quotient ring A/I.*

Again, there is a product representation

$$\zeta_C(s) = \prod_{\wp} \frac{1}{(1 - (N\wp)^{-s})}, \quad \Re(s) \gg 0, \tag{2.9}$$

where the product is taken over the prime ideals $\wp$ of A, but A being a Dedekind domain, these are maximal. Thus $A/\wp$ is a finite field extension $\mathbf{F}_q$, $q = p^f$, of $\mathbf{F}_p$ and $N\wp = p^{\deg(\wp)}$, where $\deg(\wp)$ is now called the **degree** of the **closed point** x of C, corresponding to $\wp \in \mathrm{Spec}(A)$.

For an arbitrary **non-singular algebraic variety** X over the finite field $\mathbf{F}_q$, $q = p^f$, A. Weil (1949) used this product formula as the defining expression for the **ζ-function** $\zeta_X(s)$ of X. The definition becomes:

Definition 2.2.2 *For a non-singular algebraic variety $X/\mathbf{F}_q$,*

$$\zeta_X(s) = \prod_{x\in|X|} \frac{1}{(1-q^{-s.\deg(x)})} = \prod_{x\in|X|} \frac{1}{(1-N(x)^{-s})},$$

where x runs over the set of closed points $|X|$ of X, $N(x)$ denotes the number of elements of the residue field $k(x)$ of X in x, and $\deg(x) = [k(x) : \mathbf{F}_q]$.

With $T = q^{-s}$ and $Z(X/\mathbf{F}_q, T) = \zeta_X(s)$, it is not difficult to prove

$$Z(X/\mathbf{F}_q, T) = \exp\left(\sum_{n=1}^{\infty} \frac{T^n}{n} N_n\right), \tag{2.10}$$

where $N_n = \sum_{\deg(x)|n} \deg(x)$ is just the number of points of X with coordinates in $\mathbf{F}_{q^n}$, or in other words, the cardinality of $X(\mathbf{F}_{q^n})$. In formula:

$$N_n = \#(X(\mathbf{F}_{q^n})). \tag{2.11}$$

These ζ- (or Z-) functions were the starting point for the formulation of the famous Weil Conjectures for varieties over finite fields. They gave rise to a totally new set-up of algebraic geometry, the language of schemes and étale cohomology (in particular, ℓ-adic cohomology), by a whole group of prominent mathematicians under the leadership of A. Grothendieck, at the Institut des Hautes Études Scientifiques (IHES) at Bures-sur-Yvette near Paris. After about 25 years of intensive work during which important partial results were obtained (mainly by Grothendieck), it was P. Deligne who, in 1973, gave an astonishing proof of the last remaining Weil Conjecture, the so-called Riemann-Weil Hypothesis for varieties over finite fields.

For a non-singular projective curve C of genus g defined over $\mathbf{F}_q$, $q = p^f$, Weil (1949) was able to show that his Z-function $Z(C/\mathbf{F}_q, T)$ could be written as a rational function

$$Z(C/\mathbf{F}_q, T) = \frac{P(C/\mathbf{F}_q, T)}{(1-T)(1-qT)}, \tag{2.12}$$

with

$$P(C/\mathbf{F}_q, T) = \prod_{i=1}^{2g}(1-\alpha_i T), \quad |\alpha_i| = \sqrt{q}, \tag{2.13}$$

which implies, after the substitution $T = q^{-s}$, that all zeroes of $\zeta_C(s)$ lie on the line $\Re(s) = \frac{1}{2}$. This is just the Riemann Hypothesis for the curve $C/\mathbf{F}_q$. Moreover, $P(C/\mathbf{F}_q) \in \mathbf{Z}[T]$. For the number of $\mathbf{F}_{q^n}$-points of C one deduces from the preceding results

$$N_n = 1 + q^n - \sum_{i=1}^{2g} \alpha_i^n . \tag{2.14}$$

This means that $N_n = \#\left(C(\mathbf{F}_{q^n})\right)$ satisfies the inequality, already conjectured by Artin and proven for elliptic curves ($g = 1$) by H. Hasse in 1934,

$$|N_n - (1 + q^n)| \leq 2g\sqrt{q^n} \tag{2.15}$$

2.3 *L*-Functions of Elliptic Curves

We return to elliptic curves over $\mathbf{Q}$. Let $E/\mathbf{Q}$ be an elliptic curve. The **global (Hasse-Weil)** L**-function** of E is defined as the product of the inverses of the polynomials P of the reductions modulo all primes p of the curve E. More precisely, one defines:

Definition 2.3.1 $L(E/\mathbf{Q}, s) = \prod_p P(E_p/\mathbf{F}_p, p^{-s})^{-1}$, $\Re(s) > \frac{3}{2}$.

Here, for primes where the reduction of $E/\mathbf{Q}$ is good, the $P(E_p/\mathbf{F}_p, p^{-s})$ are the polynomials defined in the previous section. For primes p where E has bad reduction one defines

Definition 2.3.2 *The* $P(E_p/\mathbf{F}_p, T)$ *are given by*

$$P(E_p/\mathbf{F}_p, T) = \begin{cases} 1 - T & if \quad E \ has \ semi-stable \ reduction \ with \\ & \qquad \mathbf{F}_p - rational \ tangents \ at \ p\,, \\ 1 + T & if \quad E \ has \ semi-stable \ reduction \ with \\ & \qquad \mathbf{F}_p - irrational \ tangents \ at \ p\,, \\ 1 & if \quad E \ has \ unstable \ reduction \ at \ p\,. \end{cases}$$

In short, one can say that

$$L(E/\mathbf{Q}, s) = \prod_p \frac{1}{(1 - a_p p^{-s} + \varepsilon(p).p^{1-2s})}, \quad \Re(s) > \frac{3}{2}, \tag{2.16}$$

where we have put

$$\varepsilon(p) = \begin{cases} 1, & E \text{ has good reduction at } p \\ 0, & E \text{ has bad reduction at } p, \end{cases}$$

and $a_p = 1+p-\#(E_p(\mathbf{F}_p))$ $(=1+p-N_1 = \alpha_1+\alpha_2 = \alpha_1+\bar{\alpha}_1$, whenever E_p is non-singular). In fact, α_1 and $\alpha_2 = \bar{\alpha}_1$ are the characteristic roots of the Frobenius map

$$\pi_p : E_p \longrightarrow E_p ,$$

given by $\pi_p : (x, y) \mapsto (x^p, y^p)$, regarded as an element of the ring of endomorphisms of E_p, and thus,

$$a_p = 1 + p - \#(E_p(\mathbf{F}_p)) = \mathrm{Tr}(\pi_p) ,$$

when E has good reduction at p.
The following examples are taken from [Ta3].

Example 2.3.1 $E/\mathbf{Q} : y^2 + y = x^3 - x^2$ has $\Delta(E) = -11$ and after reduction mod 11 one gets the singular curve over $\mathbf{F}_{11}$:

$$E_{11}/\mathbf{F}_{11} : (y-5)^2 = (x-7)(x-8)^2 ,$$

which has a **node** at $(8,5) \in \mathbf{F}_{11} \times \mathbf{F}_{11}$. For the tangent lines $y = \alpha x + \beta$ at $(8,5)$, one finds $\alpha = 1$ or $\alpha = 10$ (in $\mathbf{F}_{11}$), so E has **semi-stable reduction** at $p = 11$, with **rational tangents** at $(8,5)$. Counting points of $E_{11}/\mathbf{F}_{11}$ gives $\#(E_{11}(\mathbf{F}_{11})) = 11$ (including the point at infinity), thus $a_{11} = 1 + 11 - \#(E_{11}(\mathbf{F}_{11})) = 1$ and $L(E/\mathbf{Q})$ becomes

$$L(E/\mathbf{Q}, s) = \frac{1}{1 - 11^{-s}} \cdot \prod_{p \neq 11} \frac{1}{(1 - a_p p^{-s} + p^{1-2s})} .$$

Example 2.3.2 $E/\mathbf{Q} : y^2 + y = x^3 - x$ has $\Delta(E) = 37$ and after reduction mod 37 one finds:

$$E_{37}/\mathbf{F}_{37} : (y-18)^2 = (x-27)(x-5)^2 ,$$

which has a **node** at $(5,18) \in \mathbf{F}_{37} \times \mathbf{F}_{37}$. For the tangents $y = \alpha x + \beta$ at $(5,18)$ one finds $\alpha^2 = -22$ in $\mathbf{F}_{37}$, but $(\frac{-22}{37}) = -1$, thus α is **irrational** over $\mathbf{F}_{37}$ (but becomes rational over the quadratic extension

$\mathbf{F}_{37^2}$). Counting $\mathbf{F}_{37}$-rational points gives $\#(E_{37}(\mathbf{F}_{37})) = 39$, thus $a_{37} = -1$ and we have:

$$L(E/\mathbf{Q}, s) = \frac{1}{1+37^{-s}} \cdot \prod_{p \neq 37} \frac{1}{(1 - a_p p^{-s} + p^{1-2s})} \,.$$

Example 2.3.3 $E/\mathbf{Q} : y^2 - y = x^3 - 7$. The projective version of this curve can be shown to equivalent to the Fermat curve $x^3 + y^3 = z^3$. One has $\Delta(E) = -3^9$ and reduction mod 3 gives:

$$E_3/\mathbf{F}_3 : (y-2)^2 = x^3 \,,$$

which has a cusp at $(0,2)$, in other words E has unstable reduction at $p = 3$. Counting points over $\mathbf{F}_3$ gives $\#(E_3(\mathbf{F}_3)) = 4$, thus $a_3 = 0$ and we get:

$$L(E/\mathbf{Q}, s) = \prod_{p \neq 3} \frac{1}{(1 - a_p p^{-s} + p^{1-2s})} \,.$$

Remark 2.3.1 It can be shown that the L-function $L(E/\mathbf{Q}, s)$ can be regarded as a (generalized) Artin L-function, coming from a system of ℓ-adic representations of the absolute Galois group $\mathrm{Gal}(\bar{\mathbf{Q}}/\mathbf{Q})$ on the first cohomology space $H^1(E \otimes \bar{\mathbf{Q}}, \mathbf{Q}_\ell)$. The a_p are the traces of the geometric Frobenius $F_p \in \mathrm{Gal}(\bar{\mathbf{Q}}/\mathbf{Q})$ acting on the cohomology space $H^1(E_p \otimes \mathbf{F}_p, \mathbf{Q}_\ell)$, which is isomorphic with $H^1(E \otimes \bar{\mathbf{Q}}, \mathbf{Q}_\ell)$, $p \neq \ell$.

As a matter of fact, the L-function is to be considered as an object that contains global information of the curve, provided by the local factors $L_p(E/\mathbf{Q}, s) = L(E_p/\mathbf{F}_p, s)$. The infinite prime $p = \infty$, we will see, is dealt with by the insertion of a certain Γ-factor into the product expression for the L-function.

To write down a functional equation for $L(E/\mathbf{Q}, s)$ one needs the notion of the conductor which measures the badness of the reduction of E at the primes dividing the discriminant $\Delta(E)$. This conductor $N = N(E/\mathbf{Q})$ is defined as

$$N(E/\mathbf{Q}) = \prod_{p|\Delta(E)} p^{f_p} \,, \text{ with } f_p = \mathrm{ord}_p\Delta(E) + 1 - n \,, \tag{2.17}$$

where $n = 1$ if $E_p/\mathbf{F}_p$ is elliptic, whereas for singular E_p, n is a positive integer such that p exactly divides N if E has multiplicative reduction at p, and in case E has additive reduction at p, p^2 divides $\Delta(E)$ and p^2 exactly divides $\Delta(E)$ if $p > 3$. The case of a cusp in characteristic 2 or 3 must be dealt with separately.

Remark 2.3.2 An accurate definition of n can be given: n is the number of irreducible components (counted with multiplicity one) of the fibre over $\mathbf{F}_p$ of the so-called (minimal) Néron model $\mathcal{E}$ of $E/\mathbf{Q}$. This Néron model $\mathcal{E}$ of E is a regular scheme over Spec($\mathbf{Z}$), such that $E/\mathbf{Q}$ is the generic fibre of $\mathcal{E}$.

With this conductor N one may define the modified L-function

$$\Lambda(E/\mathbf{Q}, s) = N^{\frac{s}{2}} (2\pi)^{-s} \Gamma(s) L(E/\mathbf{Q}, s). \tag{2.18}$$

Then the following conjecture can be formulated:

Conjecture 2.3.1 (Hasse-Weil) *$L(E/\mathbf{Q}, s)$ can be extended to an* entire *function and $\Lambda(E/\mathbf{Q}, s)$ satisfies the functional equation*

$$\Lambda(E/\mathbf{Q}, s) = w\,\Lambda(E/\mathbf{Q}, 2 - s) \text{ with } w = \pm 1.$$

Remark 2.3.3 Functional equations of this type with Γ-factors determined by the Hodge type of the cohomology space involved in the definition of the L-function at hand, and with the $\pm$-sign generalized to 'ε-factors' (cf. [De2]) also turn up in higher dimensional cases. They are then part of a collection of generally accepted conjectures, the Standard Conjectures on L-functions of Algebraic Varieties (over number fields). These standard conjectures will be stated in Chapter 3.

2.4 Complex Multiplication and Modular Elliptic Curves

The analyticity of $L(E/\mathbf{Q}, s)$ and the functional equation have been proven in several cases. E.g. one such case occurs when E has complex multiplication. This means that the ring of endomorphisms of E, End(E), is strictly bigger than the ring $\mathbf{Z}$. In this situation End(E) can be shown to be an order in an imaginary quadratic number field. Elliptic curves with complex multiplication, CM-curves for short, bear a special relation to number theoretic questions, e.g. if the endomorphism ring of such a CM-curve is equal to the whole ring of integers of an imaginary quadratic number field K, then this field must have class number one. This follows from class field theory. So by Gauß's Class Number Problem and its solution (cf. Conjecture 1.1.2) the number field K can be only $\mathbf{Q}(\sqrt{d})$, with $d = -1, -2, -3, -7, -11, -19, -43, -67$, or -163.

Example 2.4.1 Let $E/\mathbf{Q}$ be the curve $y^2 = x^3 - x$, then E has complex multiplication by $\mathbf{Z}[\sqrt{-1}]$.

For these CM-curves the L-function $L(E/\mathbf{Q}, s)$ can be written as a so-called **Hecke L-function**, a generalization of a Dirichlet L-function. In fact, for any elliptic curve E defined over a number field F with complex multiplication by (an order in) the ring of integers $\mathcal{O}_K$ of an imaginary quadratic number field K, M. Deuring, in a series of papers written in the 1950's, proved that the L-function $L(E/F, s)$ can be expressed as a (product of such) Hecke L-function(s). We indicate briefly what is meant by this observation, cf. [Gro], [La] and [Sh].

So let E/F be an elliptic curve over the number field F such that

$$\theta : K \xrightarrow{\sim} \mathrm{End}(E) \otimes \mathbf{Q},$$

where K is an imaginary quadratic field (not necessarily of class number 1). The isomorphism θ can be taken to be **normalized**, by which we mean that for any differential form ω of the first kind (i.e. without poles) of E (which can always be taken over F) and any $\mu \in K$, the endomorphism $\theta(\mu)$ induces a map θ^* such that

$$\theta^*(\mu)\,\omega = \omega \circ \theta(\mu) = \mu\,\omega.$$

To state Deuring's result one usually distinguishes between two situations:
(*i*) $K \subset F$, and
(*ii*) $K \not\subset F$.
Assume we are in the first case. If now $\wp$[1] is a prime of F where E has good reduction $E_\wp$, i.e. there exists an elliptic curve $E_\wp$ over the

[1]We do not distinguish between a prime and the corresponding discrete valuation, etc.

discrete valuation ring Spec($\mathcal{O}_\wp$) such that $E \simeq E_\wp \times_{\mathcal{O}_\wp} F$, then, writing $k = k(\wp)$ for the residue class field of $F_\wp$, reduction of endomorphisms gives an injection

$$\theta_\wp : K \longrightarrow \mathrm{End}_k(E_\wp) \otimes \mathbf{Q}\,,$$

whose image contains the Frobenius endomorphism $\pi_\wp$, i.e. exponentiation of the coordinates by $N\wp$. Let $\alpha_\wp$ denote the unique element of K such that $\theta_\wp(\alpha_\wp) = \pi_\wp$. Applying this to the various primes $\wp$ of F where E has good reduction, the maps $\wp \mapsto \alpha_\wp$ give rise to a so-called **Hecke character** χ_E of F with values in K^*. This Hecke character χ_E is a continuous homomorphism

$$\chi_E : \mathbf{A}_F^* \longrightarrow K^*\,,$$

where $\mathbf{A}_F^*$ denote the group of idèles of F. χ_E is said to be **unramified** at the prime $\wp$ if it is trivial on the local $\wp$-units $\mathcal{O}_\wp^*$. If this is the case, one defines

$$\chi_E(\wp) = \chi_E(x)\,,$$

where $x = (\ldots, 1, x_\wp, 1, \ldots)$ is the idèle with all components equal to 1 except at $\wp$, where it is equal to $x_\wp$, an element of order 1 at $\wp$. By multiplicativity, χ_E extends to a character on the set of all ideals I of F. If $\chi_E(\mathcal{O}_\wp^*) \neq 1$, then, by continuity, there is a smallest positive integer $n_\wp$ such that $\chi_E(1 + \wp^{n_\wp}\mathcal{O}_\wp) = 1$. The **conductor** $\mathfrak{F} = \mathfrak{F}_\chi$ of $\chi = \chi_E$ is the ideal of F, defined by the product

$$\mathfrak{F} = \mathfrak{F}_\chi = \prod_\wp \wp^{n_\wp}\,,$$

and χ is ramified at $\wp$ if $\wp$ divides this conductor. $\mathfrak{F}_\chi$ is related to the conductor $N(E/F)$ of the curve E, defined in a way similar to $N(E/\mathbf{Q})$, by the formula:

$$N(E/F) = (\mathfrak{F}_\chi)^2\,.$$

The Hecke character χ_E is an isogeny invariant and it is characterized by the following three conditions:

(*i*) Ker(χ_E) $\subset \mathbf{A}_F^*$ is an open subgroup;

(*ii*) For a principal idèle $a = (\alpha)$ one has: $\chi_E(a) = N_{F/K}(\alpha)$, the norm;

(*iii*) If $a = (a_\wp)$ is an idèle with $a_\wp = 1$ at all infinite places (primes) of F and at those finite primes where E has bad reduction, then:

$$\chi_E(a) = \prod_{\text{good } \wp} \alpha_\wp^{\mathrm{ord}_\wp(a_\wp)}\ ,\ \theta_\wp(\alpha_\wp) = \pi_\wp.$$

Now, as a first result, we mention that every point of finite order of E has coordinates in the maximal abelian extension F^{ab} of F. Thus E_{tors} is defined over F^{ab}. By class field theory one has an element $(x, F) \in \mathrm{Gal}(F^{ab}/F)$, $x \in \mathbf{A}_F^*$, acting on E_{tors}. If we denote by

$$\phi : \mathbf{C}/\Lambda \longrightarrow E \otimes \mathbf{C}$$

the analytic parametrization of $E \otimes \mathbf{C}$, then ϕ induces a map, also denoted by ϕ,

$$\phi : K/\Lambda \longrightarrow E_{tors}\,.$$

The following result can be stated: For any idèle $x \in \mathbf{A}_F^*$ one has, for the composition of maps from $K/\Lambda \longrightarrow E_{tors}$,

$$(x, F)\,\phi = \phi\,\mu(x) N_{F/K}(x^{-1})\,,$$

where μ is a uniquely defined homomorphism on $\mathbf{A}_F^*$ with values in K^*. As in the case of a Hecke character μ is called **unramified** at the prime $\wp$ if μ is trivial on the local $\wp$-units, and in this case, one defines $\mu(\wp)$ as $\mu(x)$, where $x = (\ldots, 1, x_\wp, 1, \ldots)$ with $\mathrm{ord}_\wp(x_\wp) = 1$.
After all these preparations one can now define a function $\psi_{E,F}$ on $\mathbf{A}_F^*$ by

$$\psi_{E,F}(s) = \mu(x)\, N_{F/K}(x^{-1})_\infty\,,$$

where $N_{F/K}(\cdot)_\infty : F_\infty^* \longrightarrow K_\infty^*$, $F_\infty = F \otimes \mathbf{R}$ and $K_\infty = K \otimes \mathbf{R}$, is the local norm map at infinity. Then this function $\psi = \psi_{E,F}$ is trivial on F^*, so it is a **Hecke character** on the idèle class group $C_F = \mathbf{A}_F^*/F^*$ of F. Using the composition

$$C_F \xrightarrow{\psi} K_\infty^* \xrightarrow{\sim} \mathbf{C}^* \times \mathbf{C}^*\,,$$

one obtains two Hecke characters, also written ψ and $\bar{\psi}$ for F.

Definition 2.4.1 *The Hecke L-function $L(s,\psi)$ corresponding to the Hecke character ψ with conductor $\mathfrak{F} = \mathfrak{F}_\psi$ is defined as*

$$L(s,\psi) = \sum_{(I,\mathfrak{F})=1} \frac{\psi(I)}{(NI)^s} = \prod_{\wp} \frac{1}{1-\psi(\wp)N\wp^{-s})},$$

where the product is taken over all primes $\wp$ where ψ is not ramified.

A similar definition holds for $L(s,\bar{\psi})$.
For these Hecke L-functions E. Hecke (1887–1947) proved that they can be holomorphically continued to the whole complex s-plane if ψ is not the trivial character, and that, after multiplication by a factor involving Γ-functions etc. to get an expression $\Lambda(s,\psi)$, one has a functional equation of the form

$$\Lambda(s,\psi) = W(\psi)\,\Lambda(1-s,\bar{\psi}),$$

where $W(\psi)$ is a complex number of absolute value 1.
Deuring's result can now be formulated as follows:

Theorem 2.4.1 (*i*) *If $\wp$ is a prime of F where E/F has good reduction, then ψ and μ are unramified at $\wp$, and $\psi(\wp) = \mu(\wp)$. Furthermore, $\theta_\wp(\mu(\wp))$ is just the Frobenius endomorphism of $E_\wp$, over the residue class field $k(\wp)$.*

(*ii*) *The L-function $L(E/F,s)$ can be written as the product*

$$L(E/F,s) = L(s,\psi)\,L(s,\bar{\psi}).$$

We now turn to the second case, i.e. of the elliptic curve E defined over the number field F and with complex multiplication by the imaginary quadratic field K, such that $K \not\subset F$. Here one is led to consider the curve $E_M = E \times_F M$, where $M = FK$, the composite field of F and K. Then there are (at most) two primes $\wp_1$ and $\wp_2$ of M lying over the prime $\wp$ of F. Assume, for the moment, that, indeed, $\wp$ splits completely in M, so $\wp_1 \neq \wp_2$. Then one of Deuring's basic results says that, if E_M has good reduction at one of the $\wp_i$, $i = 1, 2$, then in fact E has good reduction at $\wp$. Again, one can construct $\mu(x)$ and $\psi(x) = \psi_{E,M}(x)$, where now $x \in \mathbf{A}_M^*$, as in the first case with suitable modifications. Also, the Hecke L-function $L(s,\psi)$ is defined similarly as before. Write ρ for the (conjugation) automorphism of M over F. Deuring's result for this situation can be formulated as follows:

Theorem 2.4.2 (*i*) *Let E/F have good reduction at $\wp$. Then $\wp$ is unramified in M, so $\wp$ splits completely as the product of two different primes $\wp_1$ and $\wp_2$ in M. Let $\mu(\wp_i) = \psi(\wp_i)$, $i = 1, 2$. Then*

$$\mu(\wp_1)^p = \mu(\wp_2)\,.$$

(*ii*) *Let $\pi_\wp$ be the Frobenius endomorphism of the reduction $E_\wp$ over the residue class field $m(\wp)$ of $M_\wp$. Then $\pi_\wp = \pi_{\wp_1}$. Furthermore*

$$\pi_\wp = \theta_{\wp_1}(\mu(\wp_1)) \quad \textit{and} \quad \pi_\wp^p = \theta_{\wp_2}(\mu(\wp_2))\,.$$

(*iii*) *If $\wp$ ramifies in M, thus $\wp_1 = \wp_2$, then $\pi_\wp$ is not rational, and one has:*

$$\pi_\wp^2 = \pi_{\wp_1} \quad \textit{and} \quad \pi_\wp = \pm\sqrt{-N\wp}\,.$$

(*iv*) $L(E/F, s) = L(s, \psi)$.

Example 2.4.2 (cf. [Ko]) Consider the curve $E/\mathbf{Q} : y^2 = x^3 - x$ with complex multiplication by the ring of integers $\mathbf{Z}[i] = \mathbf{Z}[\sqrt{-1}]$ of the quadratic field $\mathbf{Q}(i)$. Because the class number of $\mathbf{Q}(i)$ is one, every ideal I of $\mathbf{Z}[i]$ is principal: $I = (\alpha)$ for a suitable Gaußian integer α. The curve has bad reduction at the prime $p = 2$. For the Hecke character ψ on the ideals (α) one may take

$$\psi(\alpha) = \psi((\alpha)) = \begin{cases} i^j\alpha & \text{if } \alpha \text{ is prime to 2, where } i^j \text{ is the unique} \\ & \text{power of } i \text{ such that } i^j\alpha \equiv 1 \bmod (2+2i) \\ 0 & \text{otherwise} \end{cases}$$

Then, because every non-zero ideal $I = (\alpha)$ has four generators, namely α, $-\alpha$, $i\alpha$ and $-i\alpha$, and because $\psi(1) = \psi(-1) = \psi(i) = \psi(-i) = 1$, one finds for the Hecke L-function, thus, by Deuring's result, for $L(E/\mathbf{Q}, s)$:

$$L(E/\mathbf{Q}, s) = \frac{1}{4} \sum_{\alpha \in \mathbf{Z}[i]} \frac{\psi(\alpha)}{|\alpha|^{2s}}\,,$$

where the summation runs all over non-zero $\alpha \in \mathbf{Z}[i]$. Since this summation is over pairs of integers $a+bi \in \mathbf{Z}[i]$, one needs to look at Fourier transforms, the Poisson summation formula, and theta-functions in two variables. After some intricate analysis one arrives at the desired result on analytic continuation of $L(E/\mathbf{Q},s)$ and the functional equation $\Lambda(s)=\Lambda(2-s)$, where

$$\Lambda(s)=\left(\frac{2\sqrt{2}}{\pi}\right)^{s}\Gamma(s)\,L(E/\mathbf{Q},s)\,.$$

Remark 2.4.1 In [ST] Serre and Tate generalized Deuring's results, in a representation theoretic setting, to the case of abelian varieties defined over number fields.

Other examples of elliptic curves where the conjecture is true are provided by the **modular elliptic curves**. An elliptic curve $E/\mathbf{Q}$ is called **modular** if there exists a non-constant map $\gamma: X_0(N) \longrightarrow E$, defined over $\mathbf{Q}$, where $X_0(N)$ is the modular curve over $\mathbf{Q}$ for the congruence subgroup $\Gamma_0(N)$ of $SL(2,\mathbf{Z})$. The curve $X_0(N)$ parametrizes isomorphism classes of pairs (A, C_N), with A an elliptic curve and C_N a cyclic subgroup of A of order N. Eventually N is just the conductor of E. In a more familiar language, $E/\mathbf{Q}$ is a modular elliptic curve if it can be parametrized by modular functions $x=f(z)$, $y=g(z)$, $z\in\mathcal{H}$, the upper halfplane, for $\Gamma_0(N)$. In fact, the (modified) L-function $\Lambda(E/\mathbf{Q},s)$ can be interpreted as the Mellin transform of weight 2 cusp form for $\Gamma_0(N)$, and for these cusp forms one has a functional equation which, after taking the Mellin transform, gives the desired functional equation for $\Lambda(E/\mathbf{Q},s)$. This result follows mainly from the work of A. Weil, M. Eichler and G. Shimura.

2.5 Arithmetic of Elliptic Curves

The main problem in the theory of elliptic curves over $\mathbf{Q}$ is the determination of their **rational points**. These points play the role of the units in an algebraic number field. The rational points on the curve $E/\mathbf{Q}$ form a group because addition of points on E is given by rational expressions of the coordinates. The basic result is

Theorem 2.5.1 (Mordell-Weil) *The group $E(\mathbf{Q})$ of rational points of the elliptic curve $E/\mathbf{Q}$ is* **finitely generated**.

The group $E(\mathbf{Q})$ is usually called the **Mordell-Weil group** of E. It may consist of a torsion part $E(\mathbf{Q})_{tors}$ and a free part. The Mordell-Weil Theorem tells us that $E(\mathbf{Q})_{tors}$ is finite and that the free part has finite rank. This corresponds exactly to Dirichlet's Unit Theorem 1.3.1 in the case of number fields. In fact, $E(\mathbf{Q})_{tors}$ is not too difficult and a theorem of B. Mazur gives a list of all possibilities for $E(\mathbf{Q})_{tors}$. In particular, the order $\#(E(\mathbf{Q})_{tors})$ is at most 16.

On the other hand, the determination of the rank r of the free part of $E(\mathbf{Q})$ seems to be an impossible task, in general. It is believed that any non-negative rank may occur, but this has not yet been proven. The record seems to be a result of J. Mestre that says that rank 14 does occur.

Example 2.5.1 (cf. [Si]) $E/\mathbf{Q} : y^2 = x^3 - 43x + 166$ has $E(\mathbf{Q})_{tors} = \{(3, \pm 18), (5, \pm 16), (11, \pm 32)\} \cup \{O\}$.

Example 2.5.2 In 1982 J. Mestre proved that the curve $E/\mathbf{Q}$ with equation

$$E/\mathbf{Q} : y^2 - 246xy + 36599029y = x^3 - 89199x^2 - 19339780x - 36239244$$

has rank at least 12.

Example 2.5.3 (cf. [Me]) Mestre's record alluded to above is given by the curve $E/\mathbf{Q}$ with equation

$$E/\mathbf{Q} : y^2 + 357573631y = x^3 + 2597055x^2 - 549082x - 19608054$$

which has rank at least 14.

Example 2.5.4 That fairly large numbers can occur (as numerator or denominator) may be demonstrated by an example of A. Bremner and J. Cassels (1984). They showed that $E/\mathbf{Q} : y^2 = x^3 + 877x$ has rank 1 and that the x-coordinate of a generator of $E(\mathbf{Q})$ is:

$$x = (612776083187947368101/7884153586063900210)^2 .$$

In the theory of elliptic curves over $\mathbf{Q}$, the counterpart of Dedekind's Class Number Formula is given by the now famous **Birch & Swinnerton-Dyer Conjectures**:

Conjecture 2.5.1 (Birch & Swinnerton-Dyer) *The L-function of E, $L(E/\mathbf{Q}, s)$, has a zero at $s = 1$ whose order r is equal to the rank of the Mordell-Weil group $E(\mathbf{Q})$.*

Conjecture 2.5.2 (Birch & Swinnerton-Dyer) *One has:*

$$\lim_{s \to 1} (s-1)^{-r} L(E/\mathbf{Q}, s) = \Omega \,\#\, (Ш(E))\, R(E/\mathbf{Q}) \,\#\, (E(\mathbf{Q})_{tors})^{-2} \prod_p c_p .$$

In this last formula we have put

$$\Omega = \int_{E(\mathbf{R})} |\omega| , \quad \text{where } \omega = \frac{dx}{2y + a_1 x + a_2}$$

is the generator of the cohomology space $H^0(E(\mathbf{C}), \Omega^1)$ of holomorphic 1-forms on the complex manifold $E(\mathbf{C})$, and the integral is taken over the real points $E(\mathbf{R})$ of E. Thus Ω is the **positive real period** of ω (or twice this period when $E(\mathbf{R})$ consists of two components).

$Ш(E)$ will be the subject of the next section.

c_p denotes the index of the subgroup of $\mathbf{Q}_p$-points of E with non-singular reduction mod p in $E_p/\mathbf{F}_p$, in the group $E(\mathbf{Q}_p)$ of $\mathbf{Q}_p$-points of E. For almost all p, c_p is equal to 1.

$R(E/\mathbf{Q})$ is the **elliptic regulator** and is a positive real number, defined as the volume of the lattice $E(\mathbf{Q}) \otimes \mathbf{Q}$ in $E(\mathbf{Q}) \otimes \mathbf{R}$ with respect to the metric given by the canonical Néron-Tate height function h on $E(\bar{\mathbf{Q}}) \otimes \mathbf{R}$. For a point $P = (x_p, y_p)$ in the free part of $E(\mathbf{Q})$, $h(P)$ is approximately the number of decimal digits of the numerator of x_p. By construction the global function h is the sum of local ones at all primes p, including the infinite prime ∞. With the Néron-Tate height h one defines a biadditive function $\langle\, ,\, \rangle : E(\bar{\mathbf{Q}}) \times E(\bar{\mathbf{Q}}) \longrightarrow \mathbf{R}$ by $\langle P, Q \rangle = \frac{1}{2}\left(h(P+Q) - h(P) - h(Q)\right)$.
Let $P_1, \ldots, P_r$ be the generators of $E(\mathbf{Q})/E(\mathbf{Q})_{tors}$, then, by definition,

$$R(E/\mathbf{Q}) = \det\left((\langle P_i, P_j \rangle)_{1 \le i,j \le r}\right) \tag{2.19}$$

In case $r = 0$, i.e. $E(\mathbf{Q}) = E(\mathbf{Q})_{tors}$, one defines $R(E/\mathbf{Q}) = 1$.

For later purposes we observe that $\langle P, Q \rangle$ can also be described as the **intersection number** of divisors $D(P)$ and $D(Q)$, determined by the points P and Q, on the Néron model $\mathcal{E}$ of the curve E (compactified at ∞) in the sense of Arakelov. Thus

$$\langle P,Q\rangle = \langle D(P),D(Q)\rangle = \langle D(P),D(Q)\rangle_f + \langle D(P),D(Q)\rangle_\infty,$$

with $\langle D(P),D(Q)\rangle_f = \sum \log|\mathcal{O}_{\mathcal{E},x}/(f_{1,x},f_{2,x})|$, and where the summation is over the points x of $\mathcal{E}$ such that the local ring $\mathcal{O}_{\mathcal{E},x}$ has dimension 2, and the $f_{i,x}$, $i=1,2$, are local equations for $D(P)$ and $D(Q)$ at x. Furthermore,

$$\langle D(P),D(Q)\rangle_\infty = -\log G(P,Q)\,,$$

where G is the (uniquely defined) **Green's function** on the Riemann surface (torus) $E(\mathbf{C})$, subject to the usual normalization conditions, (cf. [Ar]). For the torus $\pi : E(\mathbf{C}) \xrightarrow{\sim} \mathbf{C}/\mathbf{Z}\omega_1+\mathbf{Z}\omega_2$, one may take

$$d\mu = \frac{i\,dz\wedge d\bar{z}}{4|\omega_1\bar{\omega}_2-\omega_2\bar{\omega}_1|} \quad \text{to obtain :} \quad \int_{E(\mathbf{C})} d\mu = 1\,,$$

and Green's function is given by the **Klein function**:

$$G(P,Q) = |K(z-w)|\,, \quad \text{with } \pi(P)=z \text{ and } \pi(Q)=w.$$

The Klein function $K(u)$, $u\in\mathbf{C}$ is defined by

$$K(u) = \Delta(\omega_1,\omega_2)^{\frac{1}{2}}\, e^{-u\eta(u)/2}\,\sigma(u)\,.$$

In this last formula $\Delta(\omega_1,\omega_2)$ denotes the discriminant of the complex elliptic curve with Weierstraß equation $y^2=4x^3-g_2x-g_3$, i.e. $\Delta = g_2^3-27g_3^2$. The g_2 and g_3 are determined by the lattice $\Lambda\subset\mathbf{C}$, $\Lambda = \mathbf{Z}\omega_1+\mathbf{Z}\omega_2$ as follows :

$$g_2 = g_2(\Lambda) = 60\sum_{\omega\in\Lambda}{}'\omega^{-4} \quad\text{and}\quad g_3 = g_3(\Lambda) = 140\sum_{\omega\in\Lambda}{}'\omega^{-6}\,, \qquad (2.20)$$

where the $'$ means that $\omega=0$ must be excluded in the summation.
$\sigma(u)$ denotes **Weierstraß's σ-function** (relative to the lattice Λ). It is an entire function with simple zeroes at the points $\omega\in\Lambda$. By means of this function $\sigma(u)$ one can define another one, called the **Weierstraß ζ-function** (not to be confused with Riemann's ζ-function) $\zeta(u)$ by $\zeta(u)=$

$\sigma'(u)/\sigma(u)$. This ζ-function is a quasi-periodic function with respect to Λ:

$$\zeta(u+\omega) = \zeta(u) + \eta(\omega)\,, \quad \omega \in \Lambda\,.$$

Let $\eta_i = \eta(\omega_i)$, $i = 1, 2$, then one has the **Legendre relation** between the quasi-periods:

$$\omega_2\eta_1 - \omega_1\eta_2 = 2\pi i\,. \tag{2.21}$$

The function $\eta : \Lambda \longrightarrow \mathbf{C}$ can be extended to an $\mathbf{R}$-linear function on the whole complex plane, $\eta : \mathbf{C} \longrightarrow \mathbf{C}$. This implies in particular that the Klein function satisfies the relation $|K(u+\omega)| = |K(u)|$ for all $\omega \in \Lambda$, hence the real analyticity of the Green's function

$$G(P,Q) = |K(z-w)|$$

off the diagonal.

An algorithm and examples of $H(P)$ and $R(E/\mathbf{Q})$, especially for the Bremner-Cassels type curves and modular elliptic curves, can be found in [TZ].

2.6 The Tate-Shafarevich Group

Finally we arrive at the order $\#(Ш(E))$ of the **Tate-Shafarevich group** $Ш(E)$. This group measures the non-validity of the so-called **Hasse principle** for the curve $E/\mathbf{Q}$. For rational curves (conics) C over $\mathbf{Q}$, this principle says that C has a $\mathbf{Q}$–rational point whenever C has a real point and a $\mathbf{Q}_p$-valued point for every prime number p. So, the Hasse principle is a **local** vs. **global principle**. Unfortunately the Hasse principle is not valid for curves of higher genus, in general. E. Selmer (1951) gave one of the first counter-examples to the Hasse principle. He showed that for all primes p the curve $\Gamma : 3x^3 + 4y^3 + 5z^3 = 0$ has a $\mathbf{Q}_p$-point and, obviously Γ has real points, but there exist no $\mathbf{Q}$–rational point on Γ. There are many more examples. In this same spirit, we can't conclude about Fermat's Conjecture!

As always, let E be an elliptic curve defined over $\mathbf{Q}$. To define the Tate-Shafarevich group, one looks at the set of E-**torsors**. An E-**torsor** is a smooth curve C defined over $\mathbf{Q}$, that comes equipped with a simply transitive E-action. The trivial E-torsor is E itself, with the action given by translation. Two E-torsors C_1 and C_2 are equivalent if there is an isomorphism $\alpha : C_1 \xrightarrow{\sim} C_2$, compatible with the E-action. Any E-torsor equivalent to E is called **trivial**. The equivalence classes of E-torsors form the **Weil-Châtelet** set of $E/\mathbf{Q}$. An E-torsor C is in the trivial class if and only if C contains a rational point. As a matter of fact, for an E-torsor $C/\mathbf{Q}$, E is the Jacobian of C.

Now let $C/\mathbf{Q}$ be an E-torsor and choose any point $x_0 \in C$. Then one can associate to $C/\mathbf{Q}$ a Galois cocycle which maps an element $\sigma \in \mathrm{Gal}(\bar{\mathbf{Q}}/\mathbf{Q})$ to the uniquely defined point $P \in E$ which, under the action of E on C, maps x_0 to $\sigma(x_0)$. Then it is not hard to show that two equivalent E-torsors give rise to two cocycles differing by a coboundary. This map from the **Weil-Châtelet** set $WC(E/\mathbf{Q})$ to the Galois cohomology group $H^1(\mathrm{Gal}(\bar{\mathbf{Q}}/\mathbf{Q}), E(\bar{\mathbf{Q}}))$ can be shown to be bijective, thus giving a **group structure** on $WC(E/\mathbf{Q})$. From now on we call $WC(E/\mathbf{Q})$ the **Weil-Châtelet group** of $E/\mathbf{Q}$.

A completely analogous construction can be made over the reals $\mathbf{R}$ or over the local fields $\mathbf{Q}_p$, thus giving **Weil-Châtelet groups** $WC(E/\mathbf{R})$ and $WC(E/\mathbf{Q}_p)$. To abbreviate notation we will write $\mathbf{Q}_\infty$ for $\mathbf{R}$, thus including the infinite prime $p = \infty$. The embeddings of $\mathbf{Q}$ into its p-adic completions $\mathbf{Q}_p$, $p \leq \infty$, induce a map

$$WC(E/\mathbf{Q}) \longrightarrow \prod_{p \leq \infty} WC(E/\mathbf{Q}_p)\,.$$

Definition 2.6.1 *For an elliptic curve $E/\mathbf{Q}$ the* **Tate-Shafarevich group** *$Ш(E)$ is defined as follows:*

$$Ш(E) = \mathrm{Ker}\Big\{ WC(E/\mathbf{Q}) \longrightarrow \prod_{p \leq \infty} WC(E/\mathbf{Q}_p) \Big\}\,.$$

In this way one may think of $Ш(E)$ as the group of E-torsors which have points defined over each p-adic field, including the reals. $Ш(E)$ is known to be a commutative torsion group, but otherwise it has been a mystery ever since its invention.

Conjecture 2.6.1 *$Ш(E)$ is* **finite**.

It seems to be very difficult to prove this (in general). Cassels (1962) proved that, whenever Ш(E) is finite, its order must be a perfect square.

Only recently K. Rubin (1986) could prove the finiteness conjecture for certain elliptic curves with complex multiplication. More precisely, he proved:

Theorem 2.6.1 (Rubin) *Let E be an elliptic curve over $\mathbf{Q}$ with complex multiplication by an order in an imaginary quadratic field K, and such that $L(E/\mathbf{Q},1) \neq 0$. Then both $E(\mathbf{Q})$ and* Ш(E) *are finite.*

The finiteness of $E(\mathbf{Q})$ has been known since 1977 by work of J. Coates and A. Wiles, (cf. [CW]).

Rubin, in his proof, extends some work of F. Thaine on cyclotomic units, i.e. units in $\mathbf{Z}[\zeta_p]$ of the form $\frac{\zeta_p^a - 1}{\zeta_p - 1}$, p an odd prime and $p \not| a$, to elliptic units. It should be remarked that Thaine's work, in this respect, goes back to a result of Kummer which says that the index of the group C of cyclotomic units in the full group U of units of $\mathbf{Z}[\zeta_p]$ is equal to the class number of the number field $\mathbf{Q}(\cos\frac{2\pi}{p})$. Now the group ring $\mathbf{Z}[\mathrm{Gal}(\mathbf{Q}(\zeta_p)/\mathbf{Q})]$ acts on the p-parts of both U/C and the class group Cl_p of $\mathbf{Q}(\cos\frac{2\pi}{p})$. Thaine's result (also proved by Mazur and Wiles) says that if an element of this group ring annihilates the p-part of U/C, then it annihilates the p-part of Cl_p. Next, let the rational prime $p \not| 6\Delta(E)$ split as $p = \pi\bar{\pi}$ in K. Adjoin to K the coordinates of the points $P \in E(\bar{K})$ such that $\pi P = O$. This gives a field L. Then L/K is cyclic of order $p-1$ and totally ramified at π. The set of points P as above is a cyclic group $E(\pi)$ of order p. In the Weierstraß $\wp$-model of $E(\mathbf{C}) = \mathbf{C}/\Lambda$, where $\Lambda \subset \mathbf{C}$ is a lattice, $E(\pi)$ corresponds to the subgroup $(\frac{1}{\pi}\Lambda)/\Lambda$ of $\mathbf{C}/\Lambda$. Let a and b be division points of $\mathbf{C}/\Lambda$. Elliptic units are then units of L of the form (up to a factor)

$$\prod_{a,b} (\wp(a) - \wp(b))^{m_{ab}}.$$

They form a subgroup $U(E)$, of index equal (up to powers of 2 and 3) to the class number $h(L)$ of L, of the group of all units $\mathcal{O}_L^*$ of L. Write

$Cl(\mathcal{O}_L)_p$ for the p-part of the class group $Cl(\mathcal{O}_L)$ of L, and $(\mathcal{O}_L^*)_p$ for the p-part of $\mathcal{O}_L^*/U(E)$. An element $\sigma \in G = \mathrm{Gal}(L/K)$ determines a p-adic character $\chi = \chi_\sigma \in \mathbf{Z}_p^*$ of order $p-1$ such that $\sigma(P) = \chi_\sigma P$, for all $P \in E(\pi)$. Now, for any $\mathbf{Z}_p[G]$-module A, we write A^χ for $\frac{1}{p-1} \sum_{\sigma \in G} \chi(\sigma)\, \sigma^{-1} A$. Then, using Thaine's ideas on cyclotomic units, Rubin proved that the order of $((\mathcal{O}_L^*/U(E))_p)^\chi$ annihilates $(Cl(\mathcal{O}_L^*)_p)^\chi$. Using this result and a theorem of Coates and Wiles, Rubin then shows that $\mathrm{Gal}(M/L)^\chi = 0$, where M is the maximal abelian p-extension of L which is unramified outside π. This implies, finally, that the p-torsion part $Ш(E)[p] = 0$ and that $E(K)$, and hence $E(\mathbf{Q})$, is finite. For details we refer to [Ru] and [Wa2].

Example 2.6.1 The curve $E/\mathbf{Q} : x^3 + y^3 = 1$ has $\#(Ш(E)) = 1$ and $\#(E(\mathbf{Q})_{tors}) = 3$. The second Birch & Swinnerton-Dyer Conjecture 2.5.2 can be verified by calculation.

In 1988 V. Kolyvagin proved the finiteness of $E(\mathbf{Q})$ and $Ш(E)$ for any modular elliptic curve E which admits a Heegner point[2] of infinite order in $E(K)$, where K is an imaginary quadratic number field in which the primes dividing the conductor of E split. Kolyvagin's theorem has been refined by Bump, Friedburg and Hoffstein such that this condition is satisfied if the sign in the functional equation (Conjecture 2.3.1) of $L(E/\mathbf{Q}, s)$ is $+1$. Finally, together with results of Gross and Zagier, the refined version of Kolyvagin's theorem leads to the following result:

Theorem 2.6.2 *For any* modular elliptic curve $E/\mathbf{Q}$, $L(E/\mathbf{Q}, 1) \neq 0$ *implies that* $E(\mathbf{Q})$ *and* $Ш(E)$ *are* finite groups.

That this theorem implies a fundamental result, not just for a subset of the total collection of elliptic curves, defined over $\mathbf{Q}$, would follow from another, long-standing conjecture, in agreement with Langlands's Philosophy. Namely, one has

Conjecture 2.6.2 (Shimura-Taniyama-Weil) *Every elliptic curve defined over* $\mathbf{Q}$ *is* modular.

---

[2]For the definition of a Heegner point, cf. Chapter 10.

Using results from the theory of modular forms, one sees that the Shimura-Taniyama-Weil Conjecture is equivalent to the Hasse–Weil Conjecture 2.3.1 for elliptic curves defined over $\mathbf{Q}$.

A very remarkable theorem, due to G. Frey, B. Mazur, K. Ribet and J-P. Serre (1987), is the following:

Theorem 2.6.3 (Frey–Mazur–Ribet–Serre) *The truth of the* Shimura-Taniyama-Weil Conjecture *implies* Fermat's Last Theorem.

The proof of this theorem, inspired by Frey, is based on the study of the elliptic curve

$$E/\mathbf{Q} : y^2 = x(x - a^\ell)(x + b^\ell)\,,\ \ell \geq 5 \text{ prime}\,, \tag{2.22}$$

where a and b belong to the triple (a, b, c) of relatively prime integers, satisfying

$$a^\ell + b^\ell + c^\ell = 0\,,$$

ordered in such a way that b is even and $a \equiv 3 \pmod 4$. This curve would have properties too miraculous to make it exist! Suspicion arises from the fact that the discriminant $\Delta(E) = \dfrac{(abc)^\ell}{256}$ (this is an integer because $\ell \geq 5$) is 'very large' compared to the conductor N, which can be shown to be equal to $\prod_{p|abc} p$. This last feature seems to be very unlikely on behalf of another conjecture, proved in the function field case, of L. Szpiro (1983), which relates the discriminant to the conductor.

For the actual proof, whose final version is due to K. Ribet, one studies the $GL(2, \mathbf{F}_\ell)$-representation of the Galois group $\mathrm{Gal}(\bar{\mathbf{Q}}/\mathbf{Q})$ given by the irreducible $\mathbf{F}_\ell[\mathrm{Gal}(\bar{\mathbf{Q}}/\mathbf{Q})]$-module $E[\ell](\bar{\mathbf{Q}})$, where $[\ell]$ denotes ℓ-torsion. The Shimura-Taniyama-Weil Conjecture tells us that this module maps injectively into the Jacobian $J(X_0(N)_{\mathbf{Q}})$ of the modular curve $X_0(N)$ and this leads to a contradiction.

Remark 2.6.1 (Cf. Remark 1.3.1)Just as in the case of Dedekind's ζ-function we emphasize the following interpretation of the Birch & Swinnerton-Dyer Conjectures 2.5.1 and 2.5.2.

Writing $L(E/\mathbf{Q}, s) = c(s-1)^r + \ldots (h.o.t)$, then one should have

(*a*) $r = \dim_{\mathbf{Q}} E(\mathbf{Q}) \otimes \mathbf{Q}$, and

(b) $c \approx_{\mathbf{Q}^*} \Omega . R(E/\mathbf{Q})$, where $\approx_{\mathbf{Q}^*}$ means that both sides of the expression are (non-zero) real numbers, differing at most by a non-zero rational number. This is weaker than the original conjectures, but it lends itself to far reaching generalization, to be discussed in subsequent chapters on Beilinson's conjectures.

2.7 Curves of Higher Genus

To end this chapter we add a few words on plane curves of genus $g \geq 2$. For a smooth projective curve of degree d one has $g = \frac{1}{2}(d-1)(d-2)$.

Example 2.7.1 $d = 1$ or $d = 2$ give $g = 0$. The curve is a line or a conic.

Example 2.7.2 $d = 3$ gives $g = 1$. The curve is an elliptic curve.

Example 2.7.3 $F_\ell : x^\ell + y^\ell = z^\ell$, $\ell > 3$, the Fermat curve, has genus $g \geq 2$.

A long-standing conjecture for smooth projective curves over $\mathbf{Q}$ (and, more generally, over number fields) was the following, due to L. Mordell (1922):

Conjecture 2.7.1 (Mordell) *A smooth projective curve, defined over* $\mathbf{Q}$, *with genus* $g \geq 2$, *has only a* finite number *of rational points.*

Many mathematicians made serious efforts to prove this conjecture, but it was not until 1983 that G. Faltings succeeded in giving a proof. Faltings's proof is very ingenious and complicated[3]. It uses deep results on abelian varieties, the higher dimensional generalizations of elliptic curves, over $\mathbf{Q}$ or over more general number fields. To any curve of genus g one associates its Jacobian, which is a g-dimensional abelian variety that carries much information of the curve itself. The main ingredient of Faltings's proof is the notion of height. Faltings constructed a height function on the moduli space of abelian varieties of dimension

[3]Recently (1990) Faltings proved a result for abelian varieties of higher dimension from which the Mordell Conjecture follows at once.

equal to the genus of the curve under consideration, and then related this height to the classical one of points in a projective space. Actually, Faltings proved much more than Mordell's conjecture, e.g. **Tate's Conjecture** on the endomorphisms of abelian varieties (again, this has to do with representations of $\mathrm{Gal}(\bar{\mathbf{Q}}/\mathbf{Q})$ on ℓ-adic cohomology, cf. Theorem 9.1.2), and **Shafarevich's Conjecture** on the finiteness of the number of isomorphism classes of abelian varieties over number fields with good reduction outside a fixed finite set of primes. In fact, Faltings showed that his results on heights implied Tate's conjecture, and then, using deep results on group schemes and the Weil conjectures, he was able to show that the Shafarevich conjecture comes out as a consequence of Tate's conjecture. Using Torelli's theorem and a result of A. Parshin, the truth of Mordell's Conjecture results.

As a corollary one gets a result about Fermat's Conjecture.

Theorem 2.7.1 *The equation $x^\ell + y^\ell = z^\ell$, $\ell > 3$, has at most a* **finite** *number of solutions in rational integers.*

Summarizing we have the following results for smooth projective curves over $\mathbf{Q}$:

$g = 0$: If the curve has a rational point, then it has **infinitely many**.

$g = 1$: The curve is an elliptic curve and the group of rational points is **finitely generated** (Mordell-Weil).

$g = 2$: The curve has only **finitely many** rational points (Mordell-Faltings).

2.8 Appendix

In this appendix we state a generalization of the Birch & Swinnerton-Dyer Conjectures for abelian varieties over $\mathbf{Q}$ and a particularly interesting formulation of the second conjecture which is completely volume theoretic. This formulation is due to S. Bloch. In the third section we discuss 1-motives in relation to Bloch's construction and as a first step to a motivic formulation of the Birch & Swinnerton-Dyer Conjectures for other kinds of varieties. This topic will show up anew in Chapter 7 as part of Beilinson's third conjecture. The presentation follows [De1] and [De4].

2.8.1 B & S-D for Abelian Varieties

Let A be an abelian variety over $\mathbf{Q}$ and let S be a finite set of primes including the prime at ∞ and such that A has non-degenerate reduction outside S, in other words, A comes from an abelian scheme A_S over the S-integers. Then, for each prime $p \notin S$, one has an abelian variety A_p over the residue field $\mathbf{F}_p$. The Z-function $Z(A_p/\mathbf{F}_p, T)$ (cf. Definition 2.2.2) contains a factor $P_p(T) = P(A_p/\mathbf{F}_p, T) = \prod_{i=1}^{2d}(1 - \alpha_{i,p}T)$, where $d = \dim_{\mathbf{Q}} A = \dim_{\mathbf{F}_p} A_p$, similarly to equation 2.13. The polynomial $P_p(T) \in \mathbf{Z}[T]$ and $|\alpha_{i,p}| = \sqrt{p}$. Also, one has the relation

$$\prod_{i=1}^{2d}(1 - \alpha_{i,p}^n) = \#\left(A_p(\mathbf{F}_{p^n})\right)),$$

where $\#\left(A_p(\mathbf{F}_{p^n}\right)$ is the number of $\mathbf{F}_{p^n}$-valued points of A_p. Analogously to the case of an elliptic curve, one defines the Hasse-Weil L-function of A as follows:

Definition 2.8.1 *The L-function of A (with respect to S) is given by*

$$L(A/\mathbf{Q}, s) = L_S(A/\mathbf{Q}, s) = \prod_{p \notin S} P_p(p^{-s})^{-1}, \quad \Re(s) \gg 0.$$

This L-function depends on the set S, but from the formulation of the conjectures one will see that this dependence is harmless.

Denote by r the rank of the group of rational points $A(\mathbf{Q})$, and write A' for the dual abelian variety. Thus $A'(\mathbf{Q}) \cong \mathrm{Pic}^0(A)$. There is the height pairing $\langle\, ,\rangle : A(\mathbf{Q}) \times A'(\mathbf{Q}) \longrightarrow \mathbf{R}$, and, analogously to the situation for elliptic curves, one defines the Tate-Shafarevich group as follows:

$$Ш(A) = \mathrm{Ker}\left\{H^1(\mathrm{Gal}(\bar{\mathbf{Q}}/\mathbf{Q}), A(\bar{\mathbf{Q}})) \longrightarrow \prod_p H^1(\mathrm{Gal}(\bar{\mathbf{Q}}_p/\mathbf{Q}_p), A(\bar{\mathbf{Q}}))\right\}. \tag{2.23}$$

Furthermore, let

$$V_\infty = \mathrm{Vol}(A(\mathbf{R})) \text{ and } V_{bad} = \mathrm{Vol}(\prod_{p \in S} A(\mathbf{Q}_p)).$$

Finally, let a_i, $i = 1, 2 \ldots, r$ (resp. a'_j, $j = 1, 2, \ldots, r$) be generators of $A(\mathbf{Q})$ (resp. $A'(\mathbf{Q})$). Then the Birch & Swinnerton-Dyer Conjectures for A take the form:

Conjecture 2.8.1 (Birch & Swinnerton-Dyer) *The L-function of A, $L(A/\mathbf{Q}, s)$, has a zero at $s = 1$, whose order is equal to the* rank *r of the Mordell-Weil group $A(\mathbf{Q})$.*

Conjecture 2.8.2 (Birch & Swinnerton-Dyer) *One has:*

$$\lim_{s \to 1} (s-1)^{-r} L(A/\mathbf{Q}, s) = \frac{\# (\text{Ш}(A)) . \det\left(\langle a_i, a'_j \rangle\right)_{1 \le i,j \le r} V_\infty V_{bad}}{\# (A(\mathbf{Q})_{tors}) \# (A'(\mathbf{Q})_{tors})}$$

Remark 2.8.1 In case $r = 0$ one defines $\det(\langle a_i, a'_j \rangle)$ to be equal to 1.

Remark 2.8.2 Just as in the case of elliptic curves, the L-function $L(A/\mathbf{Q}, s)$ can be written as a product (over $p \notin S$) of the inverses of the characteristic polynomials of the geometric Frobenius acting on the ℓ-adic cohomology spaces $H^1(A_p \otimes \bar{\mathbf{F}}_p, \mathbf{Q}_\ell)$, $\ell \neq p$.

2.8.2 Bloch's Version of B & S-D

For an abelian variety $A/\mathbf{Q}$ with dual A', S. Bloch ([Bl1]) considers extensions of the form

$$\begin{array}{ccccccccc} 0 & \longrightarrow & T(\mathbf{Q}) & \longrightarrow & G(\mathbf{Q}) & \longrightarrow & A(\mathbf{Q}) & \longrightarrow & 0 \text{ and} \\ 0 & \longrightarrow & T(\mathbf{A_Q}) & \longrightarrow & G(\mathbf{A_Q}) & \longrightarrow & \mathcal{A}^0(\mathbf{A_Q}) & \longrightarrow & 0, \end{array}$$

where $\mathbf{A_Q}$ are the $\mathbf{Q}$–adèles, T is a ($\mathbf{Q}$–split) torus with character group isomorphic with $A'(\mathbf{Q})/A'(\mathbf{Q})_{tors}$ and $\mathcal{A}^0$ is the largest open subgroup scheme with connected fibres of the Néron model $\mathcal{A}$ of A over Spec($\mathbf{Z}$). Bloch constructs a pairing $\langle\, , \rangle : A(\mathbf{Q}) \times A'(\mathbf{Q}) \longrightarrow \mathbf{R}$, which coincides with the height pairing mentioned in the previous section. Then he shows that $G(\mathbf{Q}) \subset G(\mathbf{A_Q})$ is discrete and cocompact, so it makes sense to consider the Tamagawa number (cf. [We]) of G,

$$\tau(G) = \int_{G(\mathbf{A_Q})/G(\mathbf{Q})} (\omega, (\lambda_p))$$

where ω is the left-invariant gauge form of degree n, $n = \dim(G)$, unique up to scalars, and where $(\omega, (\lambda_p))$ denotes the **Tamagawa measure** for G with suitable convergence factors (λ_p). In addition, there are $\mathrm{Gal}(\bar{\mathbf{Q}}/\mathbf{Q})$- and $\mathrm{Gal}(\bar{\mathbf{Q}}_p/\mathbf{Q}_p)$-actions on $G(\bar{\mathbf{Q}})$, so Ш(G) can be defined similarly to Ш(A) in the previous section. The group Ш(G) is conjectured to be finite. By a diagram chase one shows that Ш$(A) \cong$ Ш(G). One can also show that $A'(\mathbf{Q})_{tors} \cong \mathrm{Pic}(G)_{tors}$. Bloch's result is the following theorem.

Theorem 2.8.1 (Bloch) *The second Birch & Swinnerton-Dyer Conjecture for the abelian variety $A/\mathbf{Q}$ is equivalent to the validity of the formula*

$$\tau(G) = \frac{\#\,(\mathrm{Pic}(G)_{tors})}{\#\,(Ш(G))}.$$

Remark 2.8.3 The formula above, for arbitrary algebraic groups G over $\mathbf{Q}$ (or, more generally, a number field), is called the **Tamagawa number conjecture** by Bloch. It makes sense only if the L-function[4] of G has order zero at $s = 1$. This last condition is fulfilled by the construction of the torus T and by a suitable choice of the convergence factors (λ_p). As a matter of fact, by the very definition of $\tau(G)$ in the case of Bloch's extensions, one has

$$\lim_{s\to 1}(s-1)^{-r}\,L(A/\mathbf{Q},s) = \frac{\mathrm{Vol}(G(\mathbf{A_Q})/G(\mathbf{Q}))}{\tau(G)}, \tag{2.24}$$

where by the first Birch & Swinnerton-Dyer Conjecture or the Tamagawa number conjecture $r = \mathrm{rank}(A(\mathbf{Q})) = \mathrm{rank}(A'(\mathbf{Q}))$. Also, one may observe that $A'(\mathbf{Q}) = \mathrm{Pic}^0(A) \cong \mathrm{Ext}^1_{Alg.Groups}(A, \mathbf{G}_m)$ (according to Weil-Barsotti-Rosenlicht) and an element $\alpha \in \mathrm{Pic}^0(A)$ gives an extension of **1-motives**, (cf. [Del] and the next section):

$$0 \longrightarrow \mathbf{G}_m \longrightarrow G_\alpha \longrightarrow A \longrightarrow 0.$$

Going further one constructs in this way extensions of the form

$$0 \longrightarrow T \longrightarrow G \longrightarrow A \longrightarrow 0,$$

where T is a split torus with character group isomorphic with the free part of $A'(\mathbf{Q})$. Such an extension is part of a 1-motive.

[4]For the general definition cf. Chapter 3.

2.8.3 1-Motives, Mixed Motives and B & S-D

To obtain a generalization of the Birch & Swinnerton-Dyer Conjectures, Deligne [De2] suggested to replace the group $A(\mathbf{Q})$ by a certain extension group in a suitable category of so-called **mixed motives**. The existence and construction of this last category remains conjectural until this moment, but much work is being done and several (partial) candidates have already been presented, (cf. [De4] and Chapter 9). For abelian varieties the situation can be described more precisely in terms of 1-motives.

Definition 2.8.2 *A* 1-motive *M over an algebraically closed field k consists of a free $\mathbf{Z}$-module of finite type X, an abelian variety A and a torus T over k, together with an extension G of A by T, and a homomorphism $u : X \longrightarrow G(k)$.*

Notation : $M = [X \xrightarrow{u} G]$, where M can be considered as a complex of group schemes, placed in degrees zero and one.

There is an equivalence $M \longrightarrow T_B(M)$ between the category of 1-motives over $\mathbf{C}$ and the category of mixed Hodge structures without torsion, H, of type $\{(0,0),(0,-1),(-1,0),(-1,-1)\}$.(†).

For a 1-motive $M = [X \xrightarrow{u} G]$ one defines a mixed $\mathbf{Z}$-Hodge structure $T_B(M) = (T_{\mathbf{Z}}(M), W, F)$ without torsion, of type (†) and with $Gr^W_{-1}(T(M))$ polarizable, by means of the following commutative diagram :

$$\begin{array}{ccccccccc} 0 & \longrightarrow & H_1(G,\mathbf{Z}) & \longrightarrow & \mathrm{Lie}(G) & \xrightarrow{\exp} & G & \longrightarrow & 0 \\ & & \| & & \uparrow \alpha & & \uparrow u & & \\ 0 & \longrightarrow & H_1(G,\mathbf{Z}) & \longrightarrow & T_{\mathbf{Z}}(M) & \xrightarrow{\beta} & X & \longrightarrow & 0 . \end{array} \qquad (2.25)$$

The weight filtration is defined by

$W_{-1}(T_{\mathbf{Z}}(M)) = \mathrm{Ker}(\beta) = H_1(G,\mathbf{Z})$ and
$W_{-2}(T_{\mathbf{Z}}(M)) = H_1(T,\mathbf{Z})$ such that $Gr^W_{-1}(T_{\mathbf{Z}}(M)) = H_1(A,\mathbf{Z})$.

The Hodge filtration is given by

$F^0(T_{\mathbf{Z}}(M_{\mathbf{C}})) = \mathrm{Ker}(\alpha_{\mathbf{C}})$, where $\alpha_{\mathbf{C}} : T_{\mathbf{Z}}(M_{\mathbf{C}}) \longrightarrow \mathrm{Lie}(G)$ is the $\mathbf{C}$-linear prolongation of α,

$F^{-1}(T_{\mathbf{Z}}(M_{\mathbf{C}})) = T_{\mathbf{Z}}(M_{\mathbf{C}})$ and
$F^{1}(T_{\mathbf{Z}}(M_{\mathbf{C}})) = 0$.
One shows that this gives a mixed Hodge structure.

Sometimes, instead of **Z**–Hodge structures, one considers only **Q**–Hodge structures and one defines the category of '1-motives up to isogeny' by taking in the definition 2.8.2 for X a finite dimensional **Q**–vector space, and A and T are taken up to isogeny. In this setting one obtains a **Q**–Hodge structure $(T_{\mathbf{Q}}(M)), W, F)$ ... etc.

Assume k has characteristic zero, and let

$$\hat{T}(M) = \varprojlim T_{\mathbf{Z}/n\mathbf{Z}}(M),$$

with $T_{\mathbf{Z}/n\mathbf{Z}}(M) = \{(x,g)|u(x) = n.g\}/\{(nx, u(x)|x \in X\}$. $T_{\mathbf{Z}/n\mathbf{Z}}(M)$ and $\hat{T}(M)$ admit a weight filtration, induced from the one on $T_B(M)$. In particular, for n of the form ℓ^m, one has

$$T_\ell(M) = \varprojlim T_{\mathbf{Z}/\ell^m\mathbf{Z}}, \text{ and } Gr_{-1}^W(T_\ell) = T_\ell(A) \qquad (2.26)$$

the Tate module of A.

Finally, one can define a k-vector space $T_{DR}(M)$ with filtrations W and F. If M is a 1-motive over **C**, then

$$\hat{T}(M) = T_{\mathbf{Z}}(M) \otimes \hat{\mathbf{Z}} = \prod_\ell T_{\mathbf{Z}}(M) \otimes \mathbf{Z}_\ell \text{ and} \qquad (2.27)$$

$$(T_{DR}(M), W, F) \cong (T_{\mathbf{Z}}(M) \otimes \mathbf{C}, W, F). \qquad (2.28)$$

Thus, one sees that there are certain compatibilities between the various 'realizations' of M.

Also, $T_B(M)$ is endowed with an involution F_∞, the Frobenius at infinity, which respects W, and the $T_\ell(M)$ admit a $\mathrm{Gal}(\bar{\mathbf{Q}}/\mathbf{Q})$-action respecting W. The compatibilities also relate F_∞ to this Galois action.

Definition 2.8.3 *(i) The $T_B(M)$, $T_\ell(M)$ and $T_{DR}(M)$ are called the* Hodge, *the* ℓ-adic *and the* de Rham realizations *of the* 1*-motive* M*, respectively.*
(ii) The triple $T(M) = (T_B(M), T_\ell(M), T_{DR}(M))$ (with suitable compatibilities) is called the motive *associated to* M.

Remark 2.8.4 The definition of a 1-motive can be extended to a situation relative to a scheme S: A will be an abelian scheme over S, T is an S-split torus, G an algebraic group scheme over S, and X is a group scheme over S which is locally constant for the étale topology and defined by a free $\mathbf{Z}$–module of finite type. $T(M)$ will be a variation of mixed Hodge structures, $T_\ell(M)$ will be an ℓ-divisible group scheme over S, and $T_{DR}(M)$ is a vector bundle over S. There are compatibility morphisms between these realizations of M, just as in the case of the field $\mathbf{C}$. In particular, a 1-motive over $\mathbf{Q}$ defines a motive over $\mathbf{Q}$.

Example 2.8.1 The unit motive $\mathbf{Z}(0) = (\mathbf{Z}, \mathbf{Z}_\ell, \mathbf{Q})$.

Example 2.8.2 The Tate motive $\mathbf{Z}(1) = ((2\pi i)\mathbf{Z}, \varprojlim \mu_{\ell^n}(\mathbf{C}), \mathbf{Q})$, with $\mathbf{Z}(1)$ of pure weight -2, $\mathbf{Z}(1)_{DR} = \mathbf{Q}$ pure of Hodge filtration -1 and of Hodge type $(-1,-1)$.

Example 2.8.3 $\mathbf{Z}(n) = \mathbf{Z}(1)^{\otimes n}$, $\mathbf{Q}(n) = \mathbf{Z}(n) \otimes \mathbf{Q}$.

Example 2.8.4 If M is the 1-motive $[\mathbf{Z} \longrightarrow 0]$, then the associated motive is given by $T(M) = T([\mathbf{Z} \longrightarrow 0]) = \mathbf{Z}(0)$.

Example 2.8.5 $T([0 \longrightarrow \mathbf{G}_m]) = \mathbf{Z}(1)$.

Example 2.8.6 $T([\mathbf{Z} \longrightarrow \mathbf{G}_m])$ is an extension of $\mathbf{Z}(0)$ with $\mathbf{Z}(1)$. In [De4] Deligne conjectured the existence of a suitable category $\mathcal{MM}$ of mixed motives, with integer coefficients over S. Over a smooth $\mathbf{Z}$–scheme S one should have

$$\Gamma(S, \mathcal{O}_S^*) \xrightarrow{\sim} \mathrm{Ext}^1_{\mathcal{MM}}(\mathbf{Z}(0), \mathbf{Z}(1)).$$

If A is an abelian variety over $\mathbf{Q}$, then $T([0 \longrightarrow A]) = T(A)$, where $T(A)$ is the motive defined by the homology of A with integer coefficients, i.e.

$$T(A)_B = H_1(A(\mathbf{C}), \mathbf{Z}) \subset H_1(A(\mathbf{C}), \mathbf{Q}) = (T(A) \otimes \mathbf{Q})_B,$$

interpreted as part of the dual of the cohomology triple of Betti-Hodge, ℓ-adic and de Rham cohomology, the so-called motivic $H^1(A)$[5]. $T(A)$ may be identified with $H^1(A)(1)$, i.e. the twisted motivic cohomology.

[5]Cf. Chapter 8 for the notion of motives.

A point $a \in A$ defines a 1-motive $[\mathbf{Z} \xrightarrow{u} A]$ by $u(1) = a$. The corresponding motive $T([\mathbf{Z} \xrightarrow{u} A])$ is an extension of $\mathbf{Z}(0)$ with $T(A)$ and one conjectures the existence of a suitable category of mixed motives $\mathcal{MM}$ such that

$$A(\mathbf{Q}) \xrightarrow{\sim} \mathrm{Ext}^1_{\mathcal{MM}}(\mathbf{Z}(0), T(A)).$$

More generally, for an extension G of A by a ($\mathbf{Q}$–split) torus T, a point g of $G(\mathbf{Q})$ defines a 1-motive $[\mathbf{Z} \xrightarrow{u} G]$ by $u(1) = g$, and the corresponding motive $T([\mathbf{Z} \xrightarrow{u} G])$ is an extension of $\mathbf{Z}(0)$ by $T(G) = H_1(G)$, the motivic homology of G with integer coefficients. Again, one expects an isomorphism

$$G(\mathbf{Q}) \xrightarrow{\sim} \mathrm{Ext}^1_{\mathcal{MM}}(\mathbf{Z}(0), T(G)),$$

or, equivalently,

$$G(\mathbf{Q}) \otimes \mathbf{Q} \xrightarrow{\sim} \mathrm{Ext}^1_{\mathcal{MM}}(\mathbf{Z}(0), T(G) \otimes \mathbf{Q}).$$

For the points at infinity $G(\mathbf{R})$ one obtains mixed $\mathbf{R}$–Hodge structures endowed with a $F_\infty \in \mathrm{Gal}(\mathbf{C}/\mathbf{R})$, in other words, one should have

$$G(\mathbf{R}) \xrightarrow{\sim} \mathrm{Ext}^1_{\mathcal{H}/\mathbf{R}}(\mathbf{R}(0), T(G) \otimes \mathbf{R})$$

and this last group is a Lie group. With equation 2.24 and the notations introduced above one obtains the following formulation of the Birch & Swinnerton-Dyer Conjectures:

Conjecture 2.8.3 *There is an injection of* $\mathrm{Ext}^1_{\mathcal{MM}}(\mathbf{Q}(0), T(G) \otimes \mathbf{Q})$ *into* $\mathrm{Ext}^1_{\mathcal{H}/\mathbf{R}}(\mathbf{R}(0), T(G) \otimes \mathbf{R})$ *with* discrete *and* cocompact image *and*

$$\lim_{s \to 1} (s-1)^{-r} L(A/\mathbf{Q}, s) \approx_{\mathbf{Q}^\bullet} \mathrm{Vol}\left(\frac{\mathrm{Ext}^1_{\mathcal{H}/\mathbf{R}}(\mathbf{R}(0), T(G) \otimes \mathbf{R})}{\mathrm{Ext}^1_{\mathcal{MM}}(\mathbf{Q}(0), T(G) \otimes \mathbf{Q})}\right).$$

Chapter 3

The general formalism of L-functions, Deligne cohomology and Poincaré duality theories

In this chapter we consider smooth projective varieties defined over $\mathbf{Q}$ *and we define their L-functions. The whole formalism depends on several conjectures, suggested by the zero- and one-dimensional cases. The main ingredient of this chapter, Deligne-Beilinson cohomology, is introduced and shown to be a Poincaré duality theory. Such a (co)homology theory has the right properties to admit a formalism of characteristic classes which will generalize the classical regulator. This will be further explained in the next chapter.*

3.1 The Standard Conjectures

Let X be a smooth projective variety defined over $\mathbf{Q}$. X will have good reductions $X_p/\mathbf{F}_p$ at almost all primes p. For all integers i, $0 \leq i \leq 2\dim(X)$, one has an action of the geometric Frobenius F_p^i on the ℓ-adic étale cohomology space $H^i(X \otimes \bar{\mathbf{Q}}, \mathbf{Q}_\ell)$, induced from the Frobenius $Fr_p : X_p \longrightarrow X_p$, where $Fr_p(x) = x^p$. This follows from Grothendieck's specialization theorem, which gives an isomorphisms

$H^i(X\otimes\bar{\mathbf{Q}},\mathbf{Q}_\ell)\xrightarrow{\sim}H^i(X_p\otimes\bar{\mathbf{F}}_p,\mathbf{Q}_\ell)$. From now on we fix the integer i and write F_p for F_p^i, and we consider the F_p's as elements of $\mathrm{Gal}(\bar{\mathbf{Q}}/\mathbf{Q})$ acting on $H^i(X\otimes\bar{\mathbf{Q}},\mathbf{Q}_\ell)$. Write $P_p(T)$ for the characteristic polynomial of F_p acting on $H^i(X\otimes\bar{\mathbf{Q}},\mathbf{Q}_\ell)$ and assume that X has good reduction outside the finite set of primes $S\cup\{\ell\}$. Define

$$L_p(X,s)=P_p(p^{-s})^{-1}=\frac{1}{\det(1-F_p.p^{-s};H^i(X\otimes\bar{\mathbf{Q}},\mathbf{Q}_\ell))},$$

and

$$L_S(X,s)=\prod_{p\notin S\cup\{\ell\}}L_p(X,s)\,,\quad \Re(s)>\frac{i}{2}+1\,.$$

For $p\in S$, $p\neq\ell$, one defines

$$L_p(X,s)=\frac{1}{\det(1-F_p.p^{-s};H^i(X\otimes\bar{\mathbf{Q}},\mathbf{Q}_\ell)^{I_p})},$$

where $I_p\subset D_p\subset Gal(\bar{\mathbf{Q}}/\mathbf{Q})$ is the inertia group at p; $F_p\in D_p$. Using these definitions for all p, $p\neq\ell$, one finally makes the following

Definition 3.1.1 *The* (i^{th}) *L-function of X is*

$$L(X,s)=\prod_p L_p(X,s)\,,\quad \Re(s)>\frac{i}{2}+1\,.$$

To state a functional equation, we still need a factor at infinity $L_\infty(X,s)$, which will be given by Γ-factors, and a positive constant A, which accounts for the ramification.

$L_\infty(X,s)$ is defined by means of the Hodge structure of the cohomology of the complex manifold $X(\mathbf{C})$,

$$H^i(X(\mathbf{C}),\mathbf{Q})\otimes\mathbf{C}\cong H^i(X(\mathbf{C}),\mathbf{C})\cong\bigoplus_{\substack{p,q\geq 0\\ p+q=i}}H^{p,q}(X)\,.$$

Complex conjugation on $X(\mathbf{C})$ induces an involution F_∞ on the space $H^i(X(\mathbf{C}),\mathbf{C})$ such that $F_\infty(H^{p,q}(X))=H^{q,p}(X)$.
Write $h^{p,q}=\dim_{\mathbf{C}}H^{p,q}(X)$, and if $i=2k$, $H^{k,k}=H^{k,+}\oplus H^{k,-}$, where $H^{k,+}$ is short for

$$H^{k,+}(X)=\left\{a\in H^{k,k}(X)|F_\infty(a)=(-1)^k a\right\}$$

and similarly, write $H^{k,-}$ for

$$H^{k,-}(X) = \left\{a \in H^{k,k}(X) | F_\infty(a) = (-1)^{k+1} a\right\} .$$

Also, let $h^{k,\pm} = \dim_{\mathbf{C}} H^{k,\pm}$. Thus, $h^{k,k} = h^{k,+} + h^{k,-}$ and $H^{k,k}(X)$ is invariant under F_∞. Then one defines

$$L_\infty(X,s) = \prod_{p<q} \Gamma_{\mathbf{C}}(s-p)^{h^{p,q}} \prod_k \Gamma_{\mathbf{R}}(s-k)^{h^{k,+}} \Gamma_{\mathbf{R}}(s-k+1)^{h^{k,-}} . \quad (3.1)$$

Here we have put $\Gamma_{\mathbf{C}}(z) = 2(2\pi)^{-z}\Gamma(z)$ and $\Gamma_{\mathbf{R}}(z) = \pi^{-z/2}\Gamma(\frac{z}{2})$. In this way one verifies that $\Gamma_{\mathbf{C}}(z) = \Gamma_{\mathbf{R}}(z)\Gamma_{\mathbf{R}}(z+1)$.

Finally, one defines a positive integer A, depending on the primes where X has bad reduction, by introducing a (generalized) conductor. For details we refer to [Se1]. Combining all the above notions, one defines

$$\Lambda(X/\mathbf{Q}, s) = A^{s/2} L_\infty(X,s) L(X,s) ,$$

and the conjectural functional equation will be

$$\Lambda(X/\mathbf{Q}, s) = w \Lambda(X/\mathbf{Q}, i+1-s), \quad w = \pm 1. \quad (3.2)$$

For the factors $L_p(X,s)$ and the Euler product for $L(X,s)$ one assumes the following Standard Conjectures.

Conjecture 3.1.1 (Standard Conjectures on L-functions)

(i) $(L_p(X,s))^{-1} \in \mathbf{Z}[p^{-s}]$, *is independent of the prime* ℓ.

(ii) The Euler product $L(X,s) = \prod_p L_p(X,s)$ *converges absolutely for* $\Re(s) > \frac{i}{2}+1$, *and does not vanish in this region.*

(iii) $L(X,s)$ *admits a meromorphic continuation to the whole complex plane with at most a pole at* $s = \frac{i}{2}+1$ *when* i *is even. In particular,* $L(X,s)$ *extends to an entire function when* i *is odd.*

(iv) $L(X, \frac{i}{2}+1) \neq 0$.

(v) The functional equation $\Lambda(X/\mathbf{Q}, s) = \pm\Lambda(X/\mathbf{Q}, i+1-s)$ *holds.*

Remark 3.1.1 It follows from Deligne's proof of the Weil Conjectures that *(i)* is true for all p where X has good reduction. For p where X has bad reduction one has the result, also due to Deligne [De3], that the eigenvalues of F_p are of absolute value $p^{m/2}$, m an integer $\leq i$.

Remark 3.1.2 The s-values with $\frac{i}{2} < \Re(s) < \frac{i}{2} + 1$ form the critical **strip**, and the functional equation does not give information on the middle of this strip when the $+$-sign holds. In case the $-$-sign holds one should have $L(X/\mathbf{Q}, \frac{i+1}{2}) = 0$.

3.2 Deligne-Beilinson Cohomology

Using the isomorphism $H^i(X(\mathbf{C}), \mathbf{Q}) \otimes \mathbf{C} \xrightarrow{\sim} H^i_{DR}(X)$ between the complexified Betti cohomology and the de Rham cohomology of the complex manifold $X(\mathbf{C})$, one has the Hodge filtration on $H^i_{DR}(X)$,

$$F^p = F^p H_{DR}(X) = \bigoplus_{\substack{p' \geq p \\ p'+q=i}} H^{p',q}(X) = \operatorname{Im}\left(\mathbf{H}^i(\Omega^\bullet_{\geq p}) \xrightarrow{\alpha} \mathbf{H}^i(\Omega^\bullet)\right),$$

where $\mathbf{H}^i$ denotes the ith hypercohomology space and α is the map on the hypercohomology spaces, induced by inclusion of the complex $\Omega^\bullet_{\geq p}$ into the total complex $\Omega^\bullet$. More precisely, $\Omega^\bullet$ is the holomorphic de Rham complex

$$\begin{array}{llllllll} \Omega^\bullet & : & \mathcal{O}_{X(\mathbf{C})} & \longrightarrow & \Omega^1 & \longrightarrow & \Omega^2 & \longrightarrow \dots \quad \text{and} \\ \Omega^\bullet_{\geq p} & : & 0 & \longrightarrow & \Omega^p & \longrightarrow & \Omega^{p+1} & \longrightarrow \dots \quad , \end{array}$$

with Ω^p at the pth place. One can show that α is injective and has as cokernel the hypercohomology $\mathbf{H}^i(\Omega^\bullet_{<p})$ of the truncated complex

$$\Omega^\bullet_{<p} : \mathcal{O}_{X(\mathbf{C})} \longrightarrow \Omega^1 \longrightarrow \dots \longrightarrow \Omega^{p-1} \longrightarrow 0\,.$$

Write $H^i(X(\mathbf{C}), \mathbf{C})^{(-1)^k}$ for the corresponding $(-1)^k$-eigensubspace of the cohomology space $H^i(X(\mathbf{C}), \mathbf{C}) = H^i(X(\mathbf{C}), \mathbf{Q}) \otimes \mathbf{C}$ under F_∞. Using well-known properties of the Γ-function, (iii) and (v) of the Standard Conjectures give

Proposition 3.2.1 *Let $m \leq \frac{i}{2}$ and $n = i + 1 - m$, then*

(i) $\operatorname{ord}_{s=m} L(X,s) - \operatorname{ord}_{s=m+1} L(X,s) =$

$$= \dim_{\mathbf{C}} H^i(X(\mathbf{C}), \mathbf{C})^{(-1)^{n-1}} - \dim_{\mathbf{C}} F^n H^i_{DR}(X), \text{ if } m = \tfrac{i}{2}.$$

(ii) $\operatorname{ord}_{s=m} L(X,s) = \dim_{\mathbf{C}} H^i(X(\mathbf{C}), \mathbf{C})^{(-1)^{n-1}} - \dim_{\mathbf{C}} F^n H^i_{DR}(X),$ *if* $m < \frac{i}{2}$.

Example 3.2.1 $X = \mathrm{Spec}(\mathbf{Z})$, $i = 0$ give $L_p(X,s) = \dfrac{1}{1-p^{-s}}$ for all p, thus $L(X,s) = \prod_p \dfrac{1}{1-p^{-s}} = \zeta(s)$, the Riemann ζ-function. The functional equation is

$$\pi^{-s/2}\,\Gamma(\frac{s}{2})\,\zeta(s) = \pi^{(s-1)/2}\,\Gamma(\frac{1-s}{2})\,\zeta(1-s)\,.$$

$\zeta(s)$ has a simple pole at $s = 1$ and simple zeroes at $s = -2m$, $m = 1,2,3,\ldots$. Besides, one knows that $\zeta(0) = -\frac{1}{2}$, $\zeta(1-2m) = \dfrac{(-1)^m}{2m}\,B_m \in \mathbf{Q}^*$ and $\zeta(-2m) = \dfrac{2\pi^{2m}}{2\,(2m)!}\,B_m \approx_{\mathbf{Q}^*} (2\pi)^{2m}$, where B_m is the mth Bernoulli number.

Example 3.2.2 $X = \mathrm{Spec}(K)$, K a finite number field with $\deg(K) = [K:\mathbf{Q}] = n = r_1 + 2r_2$, $i = 0$, and $L(X,s)$ is the Artin L-function of K, i.e. Dedekind's ζ-function $\zeta_K(s)$. The functional equation is

$$A^s\,\Gamma(\frac{s}{2})^{r_1}\,\Gamma(s)^{r_2}\,\zeta_K(s) = A^{1-s}\,\Gamma(\frac{1-s}{2})^{r_1}\,\Gamma(1-s)^{r_2}\,\zeta_K(1-s)$$

where $A = 2^{-r_2}\,\pi^{-n/2}\,|D(K)|^{1/2}\,\zeta_K(s)$ has a simple pole at $s = 1$ with residue given by the Class Number Formula. Moreover, $\zeta_K(s)$ has a zero of order $r_1 + r_2 - 1$ at $s = 0$, and $\zeta_K(s)$ has zeroes of order $r_1 + r_2$ at $s = -2m$, and of order r_2 at $s = 1 - 2m$, $m = 1,2,3,\ldots$.

Example 3.2.3 X is an elliptic curve defined over $\mathbf{Q}$, $i = 1$, and the L-function is the one defined in the previous chapter. Parts (iii) and (v) correspond to the Hasse-Weil Conjecture. Parts (i), (ii) and (iv) are verified. For $i = 0$ or $i = 2$ one has $L_p(X,s) = \dfrac{1}{1-p^{-s}}$ and $L_p(X,s) = \dfrac{1}{1-p^{1-s}}$, respectively, and we are in the familiar case of the Riemann ζ-function, where the Standard Conjectures are verified.

Remark 3.2.1 In all the examples above the real problems arise in the middle of the the critical strip. For number fields on the line $\Re(s) = \frac{i}{2}$, where one has the (generalized) Riemann Hypothesis, and for elliptic curves on the line $\Re(s) = 1$ where for $s = 1$ one has the Birch & Swinnerton-Dyer Conjectures.

The right hand side of the proposition can be given a more natural character when one introduces the Deligne-Beilinson (or simply, Deligne) cohomology $H^\bullet_\mathcal{D}$. This cohomology is defined for complex analytic manifolds X as the hypercohomology of the so-called Deligne complex $\mathbf{Z}(p)_\mathcal{D}$, $p = 0, 1, 2, \ldots$:

$$\mathbf{Z}(p)_\mathcal{D} \,:\, 0 \longrightarrow (2\pi i)^p\mathbf{Z} \longrightarrow \mathcal{O}_X \longrightarrow \Omega^1_X \longrightarrow \ldots \longrightarrow \Omega^{p-1}_X \longrightarrow 0\,,$$

with $(2\pi i)^p\mathbf{Z}$ (often denoted $\mathbf{Z}(p)$) in degree zero. In this definition the ring $\mathbf{Z}$ may be replaced by any other subring A of $\mathbf{R}$. In particular, one may consider the complex $\mathbf{R}(p)_\mathcal{D}$:

$$\mathbf{R}(p)_\mathcal{D} \,:\, 0 \longrightarrow \mathbf{R}(p) \longrightarrow \mathcal{O}_X \longrightarrow \Omega^1_X \longrightarrow \ldots \longrightarrow \Omega^{p-1}_X \longrightarrow 0\,,$$

with $\mathbf{R}(p) = (2\pi i)^p\mathbf{R}$ placed in degree zero.

Usually one writes $H^i_\mathcal{D}(X, \mathbf{R}(p))$ etc. for $\mathbf{H}^i(X, \mathbf{R}(p)_\mathcal{D})$ etc. for the Deligne(-Beilinson) cohomology. The complex $\mathbf{R}(p)_\mathcal{D}$ fits into the short exact sequence

$$0 \longrightarrow \Omega^\bullet_{<p} \longrightarrow \mathbf{R}(p)_\mathcal{D} \longrightarrow \mathbf{R}(p) \longrightarrow 0$$

and this sequence leads to several useful long exact sequences.

Proposition 3.2.2 *The following sequences are exact:*

(a) $\ldots \longrightarrow H^i_\mathcal{D}(X, \mathbf{R}(p)) \longrightarrow H^i(X(\mathbf{C}), \mathbf{R}(p)) \oplus F^pH^i_{DR}(X) \longrightarrow$
$\longrightarrow H^i(X(\mathbf{C}), \mathbf{C}) \longrightarrow H^{i+1}_\mathcal{D}(X, \mathbf{R}(p)) \longrightarrow \ldots$

(b) $\ldots \longrightarrow H^i_\mathcal{D}(X, \mathbf{R}(p)) \to H^i(X(\mathbf{C}), \mathbf{R}(p)) \to H^i(X(\mathbf{C}), \mathbf{C})/F^p \longrightarrow$
$\longrightarrow H^{i+1}_\mathcal{D}(X, \mathbf{R}(p)) \longrightarrow \ldots$

(c) $\ldots \longrightarrow H^i_\mathcal{D}(X, \mathbf{R}(p)) \longrightarrow F^pH^i_{DR}(X) \longrightarrow H^i(X(\mathbf{C}), \mathbf{R}(p-1)) \longrightarrow$
$\longrightarrow H^{i+1}_\mathcal{D}(X, \mathbf{R}(p)) \longrightarrow \ldots$

In (c) we used the identification $\mathbf{C} = \mathbf{R}(p) \oplus \mathbf{R}(p-1)$. Similar exact sequences may be derived for other rings $A \subset \mathbf{R}$.

For $i < 2p$ one can derive, by complex conjugation of the coefficients, that the map

$$H^i(X(\mathbf{C}), \mathbf{R}(p)) \longrightarrow H^i(X(\mathbf{C}), \mathbf{C})/F^p = H^i_{DR}(X)/F^p$$

is injective.

The involution F_∞ combined with complex conjugation of the coefficients (sometimes denoted $\bar{F}_\infty$) induces the so-called **de Rham conjugation** on $H^i_{DR}(X)$. The invariants under this de Rham conjugation will be denoted by $H^i_{DR}(X_{/\mathbf{R}})$.

Because the Hodge-de Rham filtration is defined over $\mathbf{R}$, it makes sense to write $F^pH^i_{DR}(X_{/\mathbf{R}})$. In the same way there is a de Rham conjugation, DR-conjugation for short, on the Deligne cohomology space $H^i_{\mathcal{D}}(X, \mathbf{R}(p))$, and we will write $H^i_{\mathcal{D}}(X_{/\mathbf{R}}, \mathbf{R}(p))$ for the DR-invariants. With these notations and letting $n = i+1-m$, $m < \frac{i+1}{2}$ one deduces from the sequence (b) above, the basic short exact sequence

$$0 \to F^nH^i_{DR}(X_{/\mathbf{R}}) \to H^i(X_{/\mathbf{R}}, \mathbf{R}(n-1)) \to H^{i+1}_{\mathcal{D}}(X_{/\mathbf{R}}, \mathbf{R}(n)) \to 0\,, \tag{3.3}$$

where we use the shorthand notation $H^i(X_{/\mathbf{R}}, \mathbf{R}(n-1))$ to abbreviate $H^i(X(\mathbf{C}), \mathbf{R}(n-1))^{(-1)^{n-1}}$. Proposition 3.2.1 becomes:

Proposition 3.2.3 *For $m \le \frac{i}{2}$ and $n = i+1-m$, one has:*
(i) $\mathrm{ord}_{s=m} L(X,s) - \mathrm{ord}_{s=m+1} L(X,s) = \dim_{\mathbf{R}} H^{i+1}_{\mathcal{D}}(X_{/\mathbf{R}}, \mathbf{R}(n))$, *for* $m = \frac{i}{2}$.
(ii) $\mathrm{ord}_{s=m} L(X,s) = \dim_{\mathbf{R}} H^{i+1}_{\mathcal{D}}(X_{/\mathbf{R}}, \mathbf{R}))$, *for* $m < \frac{i}{2}$.

Deligne cohomology is one of the main tools in Beilinson's formulation of a series of conjectures on the values of L-functions of arithmetic algebraic varieties (or, more generally, motives) at certain integer values of their arguments.

The above definition of the Deligne complex for a compact, complex manifold can be generalized to the case of arbitrary schemes of finite type over $\mathbf{R}$ or over $\mathbf{C}$. To see how this can be done, one needs the notion of the **cone complex** $\mathrm{Cone}(A^\bullet \xrightarrow{f} B^\bullet) = C^\bullet_f$ of the map $f : A^\bullet \longrightarrow B^\bullet$ between the complexes $A^\bullet$ and $B^\bullet$. One defines $C^n_f = A^{n+1} \oplus B^n$ with the differential $\delta : C^n_f \longrightarrow C^{n+1}_f$ given by

$$\delta(a,b) = (-d_A(a), f(a) + d_B(b))\,,$$

where d_A and d_B are the differentials of $A^\bullet$ and $B^\bullet$, respectively. In short, one usually writes $C^\bullet_f = A^\bullet[1] \oplus B^\bullet$, and the relation between the three complexes is reflected by the triangle

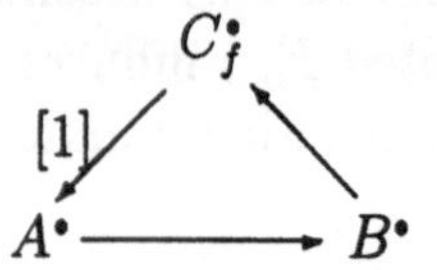

For typographical reasons such a triangle is often written as

$$\ldots \longrightarrow A^\bullet \longrightarrow B^\bullet \longrightarrow C_f^\bullet \longrightarrow A^\bullet[1] \longrightarrow \ldots .$$

The triangle gives rise to a hypercohomology long exact sequence:

$$\ldots \longrightarrow \mathbf{H}^i(A^\bullet) \longrightarrow \mathbf{H}^i(B^\bullet) \longrightarrow \mathbf{H}^i(C_f^\bullet) \longrightarrow \mathbf{H}^{i+1}(A^\bullet) \longrightarrow \ldots$$

Returning to Deligne cohomology of the compact complex manifold X, one may notice that the Deligne complex $A(p)_{\mathcal{D}}$, where A is a subring of $\mathbf{R}$, is quasi-isomorphic to the complex

$$\mathrm{Cone}(F^p\Omega_X^\bullet \oplus A(p) \xrightarrow{\varepsilon - \imath} \Omega_X^\bullet)[-1]\,,$$

where ε is the embedding $F^p \hookrightarrow \Omega_X^\bullet$ and $\imath$ is the composed inclusion $A \hookrightarrow \mathbf{C} \hookrightarrow \Omega_X^\bullet$. Taking the hypercohomology of this cone complex immediately gives the exact sequences (a),(b) and (c). This cone-construction can be naturally generalized to the situation of an arbitrary (possibly non-compact or singular) algebraic variety X over $\mathbf{C}$.

Assume that such an X has 'good' compactification $\bar{X}$ and let $j : X \hookrightarrow \bar{X}$ denote the embedding such that $D = \bar{X} \backslash X$ is a normal crossings divisor (NCD). Then one defines the **Deligne-Beilinson** or simply **Deligne** cohomology $H_{\mathcal{D}}^q(X, A(p))$ as the hypercohomology of the **Deligne-Beilinson** complex

$$A_{\mathcal{D}}(p) = \mathrm{Cone}(\mathrm{R}j_*A(p) \oplus F_D^p \longrightarrow \mathrm{R}j_*\Omega_X^\bullet)[-1]\,,$$

where F_D^p is the complex

$$F_D^p : 0 \longrightarrow \Omega_{\bar{X}}^p(\log D) \longrightarrow \Omega_{\bar{X}}^{p+1}(\log D) \longrightarrow \ldots$$

One can show that this is independent of the compactification.

The definition extends almost verbatim to the case of a smooth simplicial scheme $X_\bullet/\mathbf{C}$, with proper compactification $\bar{X}_\bullet$, such that $D_\bullet = \bar{X}_\bullet \backslash X_\bullet$ is again a simplicial normal crossings divisor. Finally, for arbitrary simplicial schemes one uses a proper hypercovering. Details can be found in [EV].

Example 3.2.4 For $A = \mathbf{Z}$ and X a smooth projective variety defined over $\mathbf{C}$ and considered as an analytic variety with the classical topology, one has $H^q_\mathcal{D}(X, \mathbf{Z}(0)) = H^q(X, \mathbf{Z})$ and using the quasi-isomorphism $\mathbf{Z}(1)_\mathcal{D} \xrightarrow{\sim} \mathcal{O}^*[-1]$ one obtains $H^q_\mathcal{D}(X, \mathbf{Z}(1)) = H^{q-1}(X, \mathcal{O}^*)$. One can prove that in this situation $H^1_\mathcal{D}(X, \mathbf{Z}(1))$ coincides with the group $\mathcal{O}^*_{X,alg}$ of **algebraic** invertible functions on X. This is not true for general A.

Example 3.2.5 $A = \mathbf{Z}$, $q = 2p$ and X as above give a particular interesting exact sequence:

$$0 \longrightarrow J^p(X) \longrightarrow H^{2p}_\mathcal{D}(X, \mathbf{Z}(p)) \longrightarrow H^{2p}(X, \mathbf{Z}(p)) \cap H^{p,p}(X) \longrightarrow 0,$$

where $J^p(X)$ denotes Griffiths's **intermediate Jacobian**:

$$J^p(X) = H^{2p-1}(X, \mathbf{C})/(\imath(H^{2p-1}(X, \mathbf{Z}(p))) \oplus F^p H^{2p-1}(X, \mathbf{C})).$$

Here $\imath$ is the morphism induced by the inclusion $\imath : \mathbf{Z}(p) \hookrightarrow \mathbf{C}$, and $H^{p,p}(X)$ is a shorthand notation for $\imath^{-1}(H^{p,p}(X))$.

Two interesting cases, to be used in the sequel, are given by the following examples.

Example 3.2.6 For smooth projective X, one has

$$H^{2p}_\mathcal{D}(X, \mathbf{R}(p)) = H^{2p}(X, \mathbf{R}(p)) \cap H^{p,p}(X), \text{ and}$$
$$H^{2p-1}_\mathcal{D}(X, \mathbf{R}(p-1)) = H^{2p-2}(X, \mathbf{R}(p-1)) \cap H^{p-1,p-1}(X).$$

Example 3.2.7 For X a smooth, quasi-projective variety one has

$$H^1_\mathcal{D}(X, A(1)) = \left\{ f \in H^0(\bar{X}, j_*\mathcal{O}_X/A(1)) \mid df \in H^0(\bar{X}, \Omega^1_{\bar{X}}(\log D)) \right\}.$$

One can also look at real versions of Deligne-Beilinson cohomology. Basically, one uses the natural projection

$$\pi_{p-1} : \mathbf{C} = \mathbf{R}(p) \oplus \mathbf{R}(p-1) \longrightarrow \mathbf{R}(p-1),$$

which induces a quasi-isomorphism

$$\mathrm{Cone}(\mathbf{R}(p) \longrightarrow \Omega_X^\bullet) \xrightarrow{\sim} \mathcal{S}_X^\bullet(p-1),$$

where $\mathcal{S}_X^\bullet$ denotes the complex of $\mathbf{R}$-valued $\mathcal{C}^\infty$-forms on $X(\mathbf{C})$, and, by definition, $\mathcal{S}_X^\bullet(p-1) = \mathcal{S}_X^\bullet \otimes_{\mathbf{R}} \mathbf{R}(p-1)$.

We apply this to a special case of Example 3.2.7 and we get

Example 3.2.8 Let X be a smooth affine curve defined over $\mathbf{R}$, with smooth compactification $\bar{X}$ such that $P = \bar{X} \backslash X$ is a finite set of points $\{x\}$. Then one derives the following result
$H^1_{\mathcal{D}}(X, \mathbf{R}(1)) =$

$$= \Big\{ \varepsilon \in \mathcal{C}^\infty(X, \mathbf{R}) | \varepsilon \in \mathrm{L}^1(X, \mathbf{C}),\ \frac{1}{\pi i} \bar{\partial}\partial\varepsilon = \sum_{x \in P(\mathbf{C})} \alpha_x \delta_x,\ \alpha_x \in \mathbf{R}[P]^0 \Big\}.$$

Here we use the following notations:
$\mathbf{R}[P]^0 = \{\text{degree zero divisors on } P\} \otimes \mathbf{R}$; $\varepsilon = \Re(\phi)$ for some $\phi \in H^0(X, \mathcal{O}_X/\mathbf{R}(1))$ such that $d\phi$ is a real analytic form on X with at most first order poles on $P(\mathbf{C})$; δ_x is the Dirac delta-distribution at x; $\alpha_x = \mathrm{Res}_x(d\phi)$. One has $2\partial\varepsilon = \partial\phi = d\phi$ and $\mathrm{div}(\varepsilon) = \mathrm{div}(2\partial\phi) = \mathrm{div}(d\phi) = \sum_x \alpha_x \delta_x \in \mathbf{R}[P]^0$.

There is an exact sequence that follows from Weyl's lemma on harmonic distributions

$$0 \longrightarrow \mathbf{R} \longrightarrow H^1_{\mathcal{D}}(X, \mathbf{R}) \xrightarrow{\mathrm{div}} \mathbf{R}[P]^0 \longrightarrow 0.$$

Furthermore, one has $H^2_{\mathcal{D}}(X, \mathbf{R}(2)) = H^1(X, \mathbf{R}(1))$, which is contained in the space of DR-invariant $\mathcal{C}^\infty$-1-forms on $X(\mathbf{C})$ modulo exact forms. Finally, there is a cup product

$$\wedge^2 H^1_{\mathcal{D}}(X, \mathbf{R}(1)) \xrightarrow{\cup} H^2_{\mathcal{D}}(X, \mathbf{R}(2)),$$

given by $\varepsilon \cup \varepsilon' = 2\{\varepsilon\pi_1(\partial\varepsilon') - \varepsilon'\pi_1(\partial\varepsilon)\}$.

As a last example we consider the case of a (finite) number field K. This example will be of importance when Borel's regulator map will be discussed in Chapter 5.

Example 3.2.9 Let $X = \mathrm{Spec}(K)$ and let $\mathrm{Hom}(K, \mathbf{C})$ denote the set of real and (pairs of conjugate) complex embeddings of K. Then

$$\mathrm{Spec}(K) \otimes \mathbf{C} = \prod_{\sigma \in \mathrm{Hom}(K,\mathbf{C})} \mathrm{Spec}(\mathbf{C}),$$

and one finds for the Deligne cohomology space

$$H^1_{\mathcal{D}}(X_{/\mathbf{C}}, \mathbf{R}(p)) = \bigoplus_{\sigma} H^1_{\mathcal{D}}(\mathrm{Spec}(\mathbf{C}), \mathbf{R}(p)) = \bigoplus_{\sigma} \mathbf{R}(p-1).$$

Taking DR-invariants, this leads to

$$H^1_{\mathcal{D}}(X_{/\mathbf{R}}, \mathbf{R}(p)) = \left[\bigoplus_{\sigma \in \mathrm{Hom}(K,\mathbf{C})} \mathbf{R}(p-1) \right]^{DR} = \begin{cases} \mathbf{R}^{r_1+r_2}, & p \geq 1, \text{ odd} \\ \mathbf{R}^{r_2}, & p \geq 2, \text{ even.} \end{cases} \tag{3.4}$$

The cup product of the fifth example is a special case of a general property, namely one can define, for any subring A of $\mathbf{R}$, a pairing

$$\mu : A(p)_{\mathcal{D}} \otimes A(q)_{\mathcal{D}} \longrightarrow A(p+q)_{\mathcal{D}}$$

by the formula

$$\mu(x \otimes y) = \begin{cases} x.y & , \text{ if } \deg(x) = 0 \\ x \wedge dy & , \text{ if } \deg(x) > 0 \text{ and } \deg(y) = q > 0 \\ 0 & , \text{ otherwise.} \end{cases}$$

Here we consider x and y as sections of one of the sheaves in the complexes $A(p)_{\mathcal{D}}$ and $A(q)_{\mathcal{D}}$, respectively, and deg denotes the degree in these complexes. The pairing μ turns out to be associative and commutative up to homotopy. It induces a cup product

$$H^i_{\mathcal{D}}(X, A(p)) \otimes_A H^j_{\mathcal{D}}(X, A(q)) \xrightarrow{\cup} H^{i+j}_{\mathcal{D}}(X, A(p+q)). \tag{3.5}$$

Using the projection π_p one can also consider the real version of this construction.

Though apparently defined in an analytic way, Deligne-Beilinson cohomology can be considered as a cohomology of certain complexes in the **Zariski topology**, more precisely, one can show that for a smooth algebraic variety X, there exists a complex $A(p)_{\mathcal{D},Zar}$ of sheaves in the Zariski topology on X such that for all open subvarieties X' of X one has

$$H^i_{\mathcal{D}}(X', A(p)) = \mathbf{H}^i(X'_{Zar}, A(p)_{\mathcal{D},Zar})\,.$$

where the absence of a subscript means that the variety is considered as an analytic variety, whereas the subscript 'Zar' means that the variety carries the Zariski topology.

Moreover, there are natural morphisms

$$c_0 : A \longrightarrow A(0)_{\mathcal{D},Zar} \text{ and } c_1 : \mathcal{O}^*_{X,Zar}[-1] \longrightarrow A(1)_{\mathcal{D},Zar}\,.$$

In case $A = \mathbf{Z}$, c_1 induces the isomorphism

$$c_1 : \mathcal{O}^*_{X',alg} \xrightarrow{\sim} H^1_{\mathcal{D}}(X', \mathbf{Z}(1))\,,\; X' \subset X\,,$$

mentioned in the first example. Also there is a pairing in the (derived) category of sheaves in the Zariski topology which induces on open subvarieties X' of X the pairing defined above.

For closed immersions $Y \hookrightarrow X$ one can construct Deligne-Beilinson cohomology with supports, $H^i_{\mathcal{D},Y}(X, A(p))$, in the usual way.

3.3 Deligne Homology

One can also construct a homological counterpart of Deligne-Beilinson cohomology for arbitrary schemes of finite type over $\mathbf{R}$ or $\mathbf{C}$. This will be done in several steps, just as in the cohomological case.

First, for a smooth compact, real or complex manifold X of dimension d, one introduces the sheaf of $(d+p, d+q)$-currents $'\Omega^{p,q}_\infty$. Thus a section on an open subset $U \subset X$ is a continuous linear functional on the space of C^∞-$(-p,-q)$-forms on U with compact support. To the double complex $'\Omega^{\bullet,\bullet}_\infty$ one associates in the natural way the simple complex $'\Omega^\bullet_\infty$ with filtration F^i given by

$$F^{i\prime}\Omega^n_\infty = \bigoplus_{\substack{p+q=n\\ p\geq i}} \Omega^{p,q}_\infty\,.$$

Next, let A be a subring of $\mathbf{R}$ and consider the homological complex $(C_\bullet(X, A(k)), d)$ of $\mathcal{C}^\infty$-chains with coefficients in $A(k)$. A change of index turns this homological complex into a cohomological one, written $('C^\bullet(X, A(k)), 'd)$, where $'C^i(X, A(k)) = C_{-i}(X, A(k))$ and, at the ith place, $'d_i = (-1)^i d_i$. There is a morphism

$$\varepsilon : 'C^\bullet(X, A(k)) \longrightarrow ' \Omega^\bullet_\infty ,$$

given by integration over chains.

Definition 3.3.1 *The* Deligne homology $'H^\bullet_{\mathcal{D}}(X, A(k))$ *is the cohomology of the complex*

$$\mathrm{Cone}('C^\bullet(X, A(k)) \oplus F^{i\prime}\Omega^\bullet_\infty \xrightarrow{\varepsilon - \imath} '\Omega^\bullet_\infty)[-1] .$$

In case X is a real manifold, X may be considered to be defined over $\mathbf{C}$ and to come equipped with an anti-holomorphic involution $F_\infty : X \longrightarrow X$. F_∞ induces involutions σ on sheaves over X, and the real Deligne homology is defined as the group hypercohomology

$$'H^\ell_{\mathcal{D}}(X_{/\mathbf{R}}, A(k)) = \mathbf{H}^\ell(< \sigma >; 'C^\bullet(X, A(k))) .$$

When X is smooth but non-compact, one takes a 'good' compactification $\bar{X}$, such that $D = \bar{X} \backslash X$ is a NCD, and one defines the complex

$$'\Omega^\bullet_\infty(\log D) = \Omega^\bullet_{\bar{X}}(\log D) \otimes_{\Omega^\bullet_{\bar{X}}} '\Omega^\bullet_\infty ,$$

where $\Omega^\bullet_{\bar{X}}(\log D)$ is the complex of holomorphic forms on X that have at most logarithmic singularities along D, and where $'\Omega^\bullet_\infty$ denotes the complex of currents on $\bar{X}$. Again, there is a filtration F^i given by

$$F^{i\prime}\Omega^\bullet_\infty(\log D) = F^{i+d}\Omega^\bullet_{\bar{X}}(\log D) \otimes_{\Omega^\bullet_{\bar{X}}} '\Omega^\bullet_\infty .$$

Here $\Omega^\bullet_{\bar{X}}(\log D)$ and $'\Omega^\bullet_\infty$ are considered as modules over the differential graded algebra $\Omega^\bullet_{\bar{X}}$. There is a (generally strict) inclusion

$$\Omega^\bullet_\infty(\log D) \otimes_{\Omega^\bullet_{\bar{X}}} F^{i\prime}\Omega^\bullet_\infty \hookrightarrow F^{i\prime}\Omega^\bullet_\infty(\log D) .$$

Let $'C^\bullet_D(\bar{X}, A(k))$ denote the subcomplex of $'C^\bullet(\bar{X}, A(k))$ consisting of $\mathcal{C}^\infty$-chains on $\bar{X}$ with support on D, and write $'C^\bullet(\bar{X}, X, A(k))$ for the quotient $'C^\bullet(\bar{X}, A(k))/'C^\bullet_D(\bar{X}, A(k))$. Then again, integration induces a map

$$\varepsilon : 'C^\bullet(\bar{X}, X, A(k)) \longrightarrow '\Omega^\bullet_\infty(\log D) .$$

Definition 3.3.2 *The cone complex* $'C^\bullet_{\mathcal{D}}(\bar{X}, X, A(k))$ *is the complex*

$$\mathrm{Cone}('C^\bullet(\bar{X}, X, A(k)) \oplus F^{k\prime}\Omega^\bullet_\infty(\log D) \xrightarrow{\varepsilon - \iota} {'\Omega^\bullet_\infty}(\log D))$$

and the Deligne homology *of the pair* $(\bar{X}, X)$ *is the hypercohomology of the complex* $'C^\bullet_{\mathcal{D}}(\bar{X}, X, A(k))$.

Actually, one can show that the Deligne homology does not depend on the compactification $\bar{X}$ of X, so we will write $'H^\ell_{\mathcal{D}}(X, A(k))$ for the Deligne homology of the smooth manifold X. The following proposition gives an important long exact sequence

Proposition 3.3.1 *There is a long exact homology sequence:*

$$\ldots \to {'H^\ell_{\mathcal{D}}}(X, A(k)) \to {'H^\ell_{BM}}(X, A(k)) \oplus F^{k\prime}H^\ell_{DR}(X) \xrightarrow{\varepsilon - \iota} {'H^\ell_{DR}}(X) \to$$
$$\to {'H^{\ell+1}_{\mathcal{D}}}(X, A(k)) \to \ldots,$$

where $'H^\bullet_{BM}(X, A(k)) \cong {'H^\bullet_{BM}}(X, \mathbf{Z}) \otimes A(k)$ denotes the Borel-Moore homology of X with coefficients in $A(k)$. This is isomorphic with the homology of the complex $'C^\bullet(\bar{X}, X, \mathbf{Z}) \otimes A(k)$. $'H^\bullet_{DR}(X)$ is the de Rham homology of X, i.e. the homology of the complex $'\Omega^\bullet_\infty(X)$ or, equivalently, of the complex $'\Omega^\bullet_\infty(\log D)$. For smooth connected X of dimension d, defined over $\mathbf{R}$ or $\mathbf{C}$, one has a canonical isomorphism

$$'H^\ell_{\mathcal{D}}(X, A(k)) \xrightarrow{\sim} H^{\ell+2d}_{\mathcal{D}}(X, A(k+d)).$$

This is just Poincaré duality for Deligne (co)homology.

Finally, for arbitrary or simplicial schemes (separated and of finite type) one uses simplicial resolutions and one proceeds along the lines developed in [Del].

3.4 Poincaré Duality Theories

To make the picture complete and to see how Deligne homology and cohomology match together we give a series of important properties of Deligne (co)homology which make it an appropriate theory for the general formalism of Chern classes and cycle maps. These properties are the defining properties of a so-called **(twisted) Poincaré duality theory** in the sense of S. Bloch and A. Ogus, (cf. [BO]).

Consider the category $\mathcal{V}$ of all schemes, separated and of finite type over $\mathbf{R}$ or over $\mathbf{C}$. To $\mathcal{V}$ we associate the category $\mathcal{V}^*$ whose objects are closed immersions $(Y \hookrightarrow X)$ and whose morphisms

$$(Y \hookrightarrow X) \xrightarrow{f} (Y' \hookrightarrow X')$$

are cartesian squares:

$$\begin{array}{ccc} Y & \hookrightarrow & X \\ f_Y\downarrow & \square & \downarrow f_X \\ Y' & \hookrightarrow & X'. \end{array}$$

Also, let $\mathcal{V}_*$ be the category with the same objects as $\mathcal{V}$, but whose class of morphisms consists only of proper morphisms.

Let $A(\star)_\mathcal{D}$ be a Deligne complex and, as always, write $H^\bullet_\mathcal{D}(X, A(\star))$ for Deligne cohomology. Similarly, with an extra subscript, for cohomology with supports. Thus one has a contravariant functor

$$H^* : \mathcal{V}^* \longrightarrow \mathcal{G}r\mathcal{A}b\,,$$

where $\mathcal{G}r\mathcal{A}b$ is the category of graded abelian groups. By definition

$$H^*((Y \hookrightarrow X)) = \bigoplus_i H^i_{\mathcal{D},Y}(X, A(\star))\,.$$

The following properties hold for Deligne-Beilinson cohomology:

(i) For $Z \subseteq Y \subseteq X$ there is a long exact sequence

$$\ldots \to H^i_{\mathcal{D},Z}X, A(\star)) \to H^i_{\mathcal{D},Y}(X, A(\star)) \to H^i_{\mathcal{D},Y\setminus Z}(X\setminus Z, A(\star)) \to$$
$$\to H^{i+1}_{\mathcal{D},Z}(X, A(\star)) \to \ldots.$$

(ii) One has compatibility of the long exact sequences for $Z \subseteq Y \subseteq X$ and $Z' \subseteq Y' \subseteq X'$ under morphisms $(Y \hookrightarrow X) \xrightarrow{f} (Y' \hookrightarrow X')$ and $(Z \hookrightarrow Y) \xrightarrow{g} (Z' \hookrightarrow Y')$.

(iii) If $(Z \hookrightarrow X) \in \mathcal{O}b(\mathcal{V}^*)$ and if $U \subset X$ is open and contains Z, then the map $H^i_{\mathcal{D},Z}(X, A(\star)) \longrightarrow H^i_{\mathcal{D},Z}(U, A(\star))$ is an isomorphism.

Next, writing $H_i^{\mathcal{P}}(X, A(j)$ for $'H_{\mathcal{D}}^{-i}(X, A(-j))$, one has a covariant functor

$$H_* : \mathcal{V}_* \longrightarrow \mathcal{G}r\mathcal{A}b\,,$$

defined by

$$H_*(X) = \bigoplus_i H_i^{\mathcal{P}}(X, A(\star))\,.$$

H_* defines a presheaf in the Zariski topology on each X in $\mathcal{V}$, such that for open immersions $\imath : U \hookrightarrow X$ and $\imath' : U' \hookrightarrow X'$ and proper mappings $f : U \longrightarrow U'$ and $g : X \longrightarrow X'$, and the cartesian square

$$\begin{array}{ccc} U & \overset{\imath}{\hookrightarrow} & X \\ f\downarrow & \square & \downarrow g \\ U' & \overset{\imath'}{\hookrightarrow} & X', \end{array}$$

one has a commutative diagram

$$\begin{array}{ccc} H_i^{\mathcal{P}}(U, A(\star)) & \xleftarrow{\imath^*} & H_i^{\mathcal{P}}(X, A(\star)) \\ f_! \downarrow & & \downarrow g_! \\ H_i^{\mathcal{P}}(U', A(\star)) & \xleftarrow{\imath'^*} & H_i^{\mathcal{P}}(X', A(\star)) \end{array}$$

The functor H_* has the following properties:

(i) Let $\imath : Y \hookrightarrow X$ be a closed immersion in $\mathcal{V}$, and let $\alpha : X\backslash Y \hookrightarrow X$ be the corresponding open immersion. Then there is a long exact sequence

$$\ldots \longrightarrow H_i^{\mathcal{P}}(Y, A(\star)) \xrightarrow{\imath_!} H_i^{\mathcal{P}}(X, A(\star)) \xrightarrow{\alpha^*} H_i^{\mathcal{P}}(X\backslash Y, A(\star)) \xrightarrow{\partial}$$
$$\xrightarrow{\partial} H_{i-1}^{\mathcal{P}}(Y, A(\star)) \longrightarrow \ldots.$$

(ii) For a proper morphism $f : X' \longrightarrow X$ in $\mathcal{V}$, let $Z = f(Z')$ and also let $\alpha : X'\backslash f^{-1}(Z) \hookrightarrow X'\backslash Z'$, then the following diagram is commutative:

$$\begin{array}{ccccccc} \rightarrow & H_i^{\mathcal{P}}(Z', A(\star)) & \longrightarrow & H_i^{\mathcal{P}}(X', A(\star)) & \longrightarrow & H_i^{\mathcal{P}}(X'\backslash Z', A(\star)) & \rightarrow \\ & f_* \downarrow & & f_* \downarrow & & f_*\alpha^* \downarrow & \\ \rightarrow & H_i^{\mathcal{P}}(Z, A(\star)) & \longrightarrow & H_i^{\mathcal{P}}(X, A(\star)) & \longrightarrow & H_i^{\mathcal{P}}(X\backslash Z, A(\star)) & \rightarrow \end{array}$$

A (twisted) **Poincaré duality theory with supports** combines the above cohomology theory H^* and the homology theory H_* in the following way:

(*a*) For the closed immersion $Y \hookrightarrow X$ in $\mathcal{V}$ there is a **cap product**

$$\cap : H_i^{\mathcal{D}}(X, A(r)) \otimes H^j_{\mathcal{D},Y}(X, A(s)) \longrightarrow H^{\mathcal{D}}_{i-j}(Y, A(r-s)),$$

which is a pairing of presheaves on each X in $\mathcal{V}$, and such that for each cartesian square

$$\begin{array}{ccc} Y & \hookrightarrow & X \\ f_Y\downarrow & \square & \downarrow f_X \\ Y' & \hookrightarrow & X' \end{array}$$

with proper f_X and $Y' \hookrightarrow X'$ closed, one has the **projection formula**

$$f_!(\alpha) \cap z = f_!(\alpha \cap f^!(z)),$$

for all $\alpha \in H_i^{\mathcal{D}}(X, A(r))$ and all $z \in H^j_{\mathcal{D},Y}(X', A(s))$.

(*b*) Let X in $\mathcal{V}$ be irreducible and of dimension d, then the preimage of $1 \in A$ under the isomorphism $H^{\mathcal{D}}_{2d}(X, A(d)) \xrightarrow{\sim} A$ is the **fundamental class** $\eta_X \in H^{\mathcal{D}}_{2d}(X, A(d))$ of X. For an open immersion $\alpha : X' \hookrightarrow X$ one has $\alpha^*\eta_X = \eta_{X'}$.

(*c*) (**'Poincaré duality'**) If X in $\mathcal{V}$ is smooth of dimension d and $Y \hookrightarrow X$ is a closed immersion, then the cap product $\cap$ induces an isomorphism

$$\eta_X \cap : H^{2d-i}_{\mathcal{D},Y}(X, A(d-r)) \xrightarrow{\sim} H_i^{\mathcal{D}}(Y, A(r)).$$

The class $\eta_X \in H^{\mathcal{D}}_{2d}(X, A(d)) \cong H^0_{\mathcal{D}}(X, A(0))$ corresponds to the unit in the ring structure of $H^{2*}_{\mathcal{D}}(X, A(\star))$. For a closed immersion $Y \hookrightarrow X$ of smooth schemes this gives an isomorphism

$$H^i_{\mathcal{D}}(Y, A(r)) \xrightarrow{\sim} H^{i+2p}_{\mathcal{D},Y}(X, A(r+p)),$$

where p is $\mathrm{codim}_X(Y)$.

(*d*) (**principal triviality**) If $\imath : D \hookrightarrow X$ is a smooth principal divisor in the smooth variety X, then $\imath_*\eta_D = 0$.

(*e*) If $f : X \longrightarrow Y$ is a proper morphism between equidimensional schemes, then $f_*\eta_X = (\deg(f))\eta_Y$.

(*f*) $H_i^{\mathcal{D}}(X, A(\star)) = 0$ if $i > 2\dim(X)$.

(*g*) For any X in $\mathcal{V}$ with natural map $p : \mathbf{A}^1_X \longrightarrow X$, there is an isomorphism

$$p^* : H^i_{\mathcal{D}}(X, A(\star)) \xrightarrow{\sim} H^i_{\mathcal{D}}(\mathbf{A}^1_X, A(\star)).$$

(*h*) For all X in $\mathcal{V}$ and all $n \geq 1$, the natural map $\mathbf{P}^n_X \xrightarrow{\pi} X$ and the hyperplane class $\xi \in H^2_{\mathcal{D}}(\mathbf{P}^n_X, A(1))$ induce an isomorphism

$$\sum_{k=0}^{n} \pi^*(.) \cap \xi^k : \bigoplus_{k=0}^{n} H^{\mathcal{D}}_{i-2k}(X, A(r-k)) \xrightarrow{\sim} H_i(\mathbf{P}^n_X, A(r)).$$

This implies in particular that the usual formalism of Chern classes of vector bundles in the manner of Grothendieck can be applied. Thus there is a ring homomorphism, called the Chern character

$$\mathrm{ch} : K_0(X) \longrightarrow \bigoplus_i H^{2i}_{\mathcal{D}}(X, A \otimes \mathbf{Q}(i)).$$

This will be generalized to higher algebraic K-theory in the next chapter.

(*i*) There is a natural transformation of contravariant functors on $\mathcal{V}$,

$$\mathrm{c}_1 : \mathrm{Pic}(X) \longrightarrow H^2_{\mathcal{D}}(X, A(1)).$$

(*j*) For quasi-projective X, Y in $\mathcal{V}$ there are external products

$$\times : H_i^{\mathcal{D}}(X, A(r)) \otimes H_j^{\mathcal{D}}(Y, A(s)) \longrightarrow H^{\mathcal{D}}_{i+j}(X \times Y, A(r+s)),$$

coming from the natural product

$$H^*_{\mathcal{D},X}(M, A(\bullet)) \otimes_{\mathbf{Z}} H^*_{\mathcal{D},Y}(N, A(\bullet)) \longrightarrow H^*_{\mathcal{D},X\times Y}(M \times N, A(\bullet)),$$

where $X \hookrightarrow M$ and $Y \hookrightarrow N$ are embeddings of X and Y into smooth schemes M and N, respectively.

Remark 3.4.1 It should be observed that the fundamental class of (b), applied to n-dimensional, irreducible, reduced subschemes Z of X, gives rise to a cycle map

$$\mathrm{cl}_{\mathcal{D}} : Z_n(X) \longrightarrow H^{\mathcal{D}}_{2n}(X, A(n)) .$$

Here $Z_n(X)$ denotes the group of n-dimensional cycles on X, and $\mathrm{cl}_{\mathcal{D}}([Z])$ is the image of η_Z under $H^{\mathcal{D}}_{2n}(Z, A(n)) \longrightarrow H^{\mathcal{D}}_{2n}(X, A(n))$.

When X is smooth, it follows from Poincaré duality that one has a cycle map

$$\mathrm{cl}_{\mathcal{D}} : Z^m(X) \longrightarrow H^{2m}_{\mathcal{D}}(X, A(m))$$

on the group $Z^m(X)$ of codimension m cycles, $m = \dim(X) - n$. One can show that this cycle map factors over the Chow group $CH^m(X)$ of codimension m cycles modulo rational equivalence. For a smooth projective variety X over $\mathbf{C}$ one obtains a morphism, also denoted by $\mathrm{cl}_{\mathcal{D}}$,

$$\mathrm{cl}_{\mathcal{D}} : CH^m(X) \longrightarrow H^{2m}_{\mathcal{D}}(X, \mathbf{Z}(m)) .$$

Let $A^m(X)$ and $B^m(X)$ denote the kernel and the image, respectively, of the ordinary cycle map $CH^m(X) \longrightarrow H^{2m}(X, \mathbf{Z}(m))$, then there is a commutative diagram

$$\begin{array}{ccccccc} 0 \to & A^m(X) & \longrightarrow & CH^m(X) & \longrightarrow & B^m(X) & \longrightarrow 0 \\ & \downarrow{\scriptstyle \phi_m} & & \downarrow{\scriptstyle \mathrm{cl}_{\mathcal{D}}} & & \downarrow & \\ 0 \to & J^m(X) & \to & H^{2m}_{\mathcal{D}}(X, \mathbf{Z}(m)) & \to & H^{2m}(X, \mathbf{Z}(m)) \cap H^{m,m}(X) & \to 0 \end{array} \tag{3.6}$$

Here $J^m(X)$ is Griffiths's intermediate Jacobian and ϕ_m is the Abel-Jacobi map.

As was already mentioned at the beginning of this section, Deligne (co)homology enjoys the properties (a) to (j) above. In general, one should take these properties, with the right notational modifications of course, as the defining properties of a Poincaré duality theory. Then, adapting, if necessary, the categories $\mathcal{V}$, $\mathcal{V}^*$ and $\mathcal{V}_*$ as well as the sheaf complexes in an obvious way, one has the following further examples of Poincaré duality theories.

Example 3.4.1 De Rham cohomology of varieties over a field k of characteristic zero. The Deligne complex is to be replaced by the complex $\Omega^{\bullet}_{X/k}$. The homology is the algebraic de Rham homology.

Example 3.4.2 Étale cohomology of schemes over a field k. Let ℓ be a positive integer prime to the characteristic of k, and let μ denote the étale sheaf of ℓ^{ν}-roots of unity on Spec(k) for a fixed positive integer ν. Write $\mu^n = \mu \otimes \ldots \otimes \mu$ when $n \geq 1$, and $\mu^n = \mathrm{Hom}(\mu^{-n}, \mathbf{Z}/\ell^{\nu}\mathbf{Z})$ when $n < 0$. Let $\pi : X \longrightarrow \mathrm{Spec}(k)$ be the structure map. The Deligne (co)homology groups have to be replaced by the étale (co)homology groups $H^i_Y(X, \pi^*\mu^n))$ and $H_i(Y) = H^{-i}(Y, \pi^!_Y\mu^{-n})$, ...etc.

Example 3.4.3 Integral cohomology of varieties of finite type over $\mathbf{C}$. Here the Deligne complex is replaced by $Ru_*(\underline{\mathbf{Z}})$ where $\underline{\mathbf{Z}}$ denotes the constant sheaf in the classical topology and u is the natural map of sites,

$$u : (\mathcal{S}ch/\mathbf{C})_{Class} \longrightarrow (\mathcal{S}ch/\mathbf{C})_{Zar}\,.$$

Example 3.4.4 The Chow ring of a variety X of finite type over a field. The sheaf complex will be the sheafification of the higher algebraic K-groups $K_j(U)$, $U \subset X$ open, for $j \geq 0$, and the zero sheaf if $j < 0$.

Example 3.4.5 An interesting and far reaching generalization of this last example has been proposed by S. Bloch, (cf. [Bl3]). Bloch's construction is of a very geometric nature and works for any scheme X of finite type over a field k. The construction roughly goes as follows.

Assume X is equidimensional of dimension d. Let $Z^r(X)$ denote the group of codimension r cycles on X. For $n = 0, 1, 2, \ldots$, let $\Delta^n = \mathbf{A}^n_k$ denote affine n-space over k with barycentric coordinates $(t_0, t_1, \ldots, t_n)$ such that $\sum t_i = 1$. For a subset $I \subset \{0, 1, 2, \ldots, n\}$ with $\#(I) = n-m$, one defines a face map $\partial_i : X \times \Delta^m \subset X \times \Delta^n$ by setting $t_i = 0$, $i \in I$. Next, define $Z^r(X, n)$ to be the free abelian group generated by the irreducible codimension r subvarieties of $X \times \Delta^n$ which meet all the faces of $X \times \Delta^m$, $m < n$, properly, i.e. the codimension of the irreducible components of the intersections is at least as large as the sum of the codimensions of the constituent parts. Thus $Z^r(X, n) \subset Z^r(X \times \Delta^n)$. The pullback maps give face maps $\partial_i : Z^r(X, n) \longrightarrow Z^r(X, m)$, such that, for $V \in Z^r(X,n)$, $\partial_i(V)$ equals the cycle associated to the scheme $V \cap \partial_i(X \times \Delta^m)$. One also has degeneracy maps

$$\sigma_i : X \times \Delta^{n+1} \longrightarrow X \times \Delta^n ,$$

given by

$$\sigma_i(x, t_0, \ldots, t_{n+1}) = (x, t_0, \ldots, t_{i-1}, t_i + t_{i+1}, t_{i+2}, \ldots, t_{n+1}) .$$

These σ's lead to maps, also denoted by σ_i,

$$\sigma_i : Z^r(X,n) \longrightarrow Z^r(X,n+1)$$

by pullback. Note that the σ's are flat maps. In this way one obtains a simplicial complex of abelian groups $Z^r(X,\bullet)$:

$$\ldots \longrightarrow Z^r(X,2) \begin{array}{c} \xrightarrow{\partial} \\ \xleftarrow{\sigma} \\ \xrightarrow{\partial} \\ \xleftarrow{\sigma} \\ \xrightarrow{\partial} \end{array} Z^r(X,1) \begin{array}{c} \xrightarrow{\partial} \\ \xleftarrow{\sigma} \\ \xrightarrow{\partial} \end{array} Z^r(X,0) . \tag{3.7}$$

Definition 3.4.1 Bloch's higher Chow group $CH^r(X,n)$ *is the homotopy (or, equivalently, the homology) of this complex,*

$$CH^r(X,n) = \pi_n(Z^r(X,\bullet)) = H_n(Z^r(X,\bullet)) .$$

For $n = 0$ one recovers the traditional Chow groups $CH^r(X)$.

For smooth X and $r = 1$, one gets:

$$CH^1(X,n) = \begin{cases} \mathrm{Pic}(X), & \text{if } n = 0 \\ \Gamma(X, \mathcal{O}_X^*), & \text{if } n = 1 \\ 0 & \text{if } n \geq 2. \end{cases}$$

The complex $Z^r(X,n)$ behaves well under base change: let k'/k be a Galois extension with Galois group G, and let $X' = X \otimes \mathrm{Spec}(k')$, then $Z^r(X,n) = Z^r(X',n)^G$.

For X smooth over k one may sheafify $CH^r(X,n)$ to get $\mathcal{CH}_X^r(n)$, and one obtains an important exact sequence, comparable to Gersten's sequence in algebraic K-theory:

$$0 \longrightarrow \mathcal{CH}_X^r(n) \longrightarrow \bigoplus_{x\in X^{(0)}} i_X(CH^r(\mathrm{Spec}(k(x)),n)) \longrightarrow$$
$$\longrightarrow \bigoplus_{x\in X^{(1)}} i_X(CH^{r-1}(\mathrm{Spec}(k(x)),n-1)) \longrightarrow \ldots \longrightarrow$$
$$\longrightarrow \bigoplus_{x\in X^{(n)}} i_X(CH^{r-n}(\mathrm{Spec}(k(x)),0)) \longrightarrow 0\,.$$

Here $X^{(j)}$ denotes the set of points of codimension j in X, and $i_X(A)$ means the constant sheaf with stalk A on the Zariski closure $\{\bar{x}\}$. In particular, one has $CH^r(X) \cong H^r(X, \mathcal{CH}_X^r(r))$.

Remark 3.4.2 The existence of exact sequences of this type for suitable cohomology groups of smooth algebraic varieties over some field k was the main motivation of Bloch and Ogus for the introduction of Poincaré duality theories.

Besides the special relation to K-theory (cf. the next chapter) and the general properties of a Poincaré duality theory, the groups $CH^r(X,\bullet)$ are believed to have a universal nature and there may exist an Euler-Poincaré characteristic for the complex $Z^r(X,\bullet)$, thus revealing the existence of an arithmetic index theory. More precisely, S. Bloch and C. Soulé state the following conjecture.

Conjecture 3.4.1 (Bloch-Soulé) *Let X be a scheme of finite type over* $\mathrm{Spec}(\mathbf{Z})$ *with* $\dim(X) = d$, *and with ζ-function*

$$\zeta_X(s) = \prod_{x\in|X|} \frac{1}{1 - N(x)^{-s}}\,,$$

the product being taken over the closed points $|X|$ of X. Then:
(i) after analytic continuation $\zeta_X(s)$ is meromorphic on $\mathbf{C}$;
(ii) $-\mathrm{ord}_{s=d-r}\,\zeta_X(s) = \sum_i (-1)^i \mathrm{rank}(CH^r(X,i))$.

Actually, it is unknown, at this moment, whether the $CH^r(X,i)$ have finite rank.

One may hope that the groups $CH^r(X,n)$, which are integrally defined, give precise information on the values of L-functions at certain integer values of their arguments. This would imply a sharpening of the statements in Beilinson's Conjectures to be discussed in Chapters 5, 6 and 7.

Chapter 4

Riemann-Roch, K-theory and motivic cohomology

This chapter concerns mainly algebraic K-theory as a Poincaré duality theory. Beilinson's basic idea is that this duality theory is universal and that its relation with other Poincaré duality theories is given by generalized regulator maps. An essential role is played by a Riemann-Roch Theorem in higher algebraic K-theory, due to H. Gillet.

4.1 Grothendieck-Riemann-Roch

The now classical Hirzebruch-Riemann-Roch Theorem for compact complex manifolds says (cf. [Hi]):

Theorem 4.1.1 (Hirzebruch) *Let X be a compact complex manifold and let V be a holomorphic vector bundle of rank n on X. Then one has the following formula for the* Euler-Poincaré characteristic

$$\chi(X,V) = \sum_i (-1)^i \dim_{\mathbb{C}} H^i(X,\mathcal{O}(V)) = (\mathrm{ch}(V) \cup \mathrm{td}(X))[X].$$

In this formula we have used the standard notation

$$\mathrm{ch}(V) = n + c_1(V) + \frac{1}{2}(c_1(V)^2 - 2c_2(V)) + \dots,$$

for the Chern character of V, where the $c_i(V) \in H^{2i}(X, \mathbf{Z}(i))$ denote the Chern classes of V. Thus $\mathrm{ch}(V) \in H^{2*}(X, \mathbf{Q}(\star)) = \bigoplus_i H^{2i}(X, \mathbf{Q}(i))$. Also, we have written

$$\mathrm{td}(X) = 1 + \frac{1}{2}c_1(X) + \frac{1}{12}(c_2(X) + c_1(X)^2) + \ldots \in H^{2*}(X, \mathbf{Q}(\star))$$

for the Todd class of X. Here $c_i(X) \in H^{2i}(X, \mathbf{Z}(i))$ denote the Chern classes of (the tangent bundle of) X.

Remark 4.1.1 The theorem was originally proved by Hirzebruch for projective algebraic manifolds X, but nowadays it may be derived as a special case of the Atiyah-Singer Index Theorem, which is valid in the analytic case.

The theorem was generalized and given a functorial interpretation by A. Grothendieck in the following way. For an algebraic variety (or a complex manifold) X one considers the group $F(X)$ generated by the isomorphism classes of algebraic (holomorphic) vector bundles on X. Thus an element of $F(X)$ is a formal finite sum

$$\sum_i n_i[V_i]\,,\ n \in \mathbf{Z} \text{ and } V_i \longrightarrow X \text{ is a vector bundle on } X\,.$$

$R(X)$ will denote the subgroup of elements of the form

$$[V] - [V'] - [V'']\,,\ \text{where } 0 \longrightarrow V' \longrightarrow V \longrightarrow V'' \longrightarrow 0$$

is a short exact sequence of algebraic (holomorphic) vector bundles on X. The quotient group $F(X)/R(X)$ is Grothendieck's $K_0(X)$.

Tensor products of vector bundles induce a product structure on $K_0(X)$, thus making $K_0(X)$ a commutative ring. The trivial bundle gives the unit element in $K_0(X)$.

If, instead of vector bundles, one takes coherent algebraic (analytic) sheaves, then, accordingly, defines groups $F'(X)$ and $R'(X)$, the same construction can be carried out to give the group $K_0'(X)$. The inclusion functor of the category of vector bundles on X to the category of coherent algebraic (analytic) sheaves on X is exact, i.e. it maps short exact sequences of the former category to short exact sequences of the latter. Therefore there is a natural homomorphism

$$h : K_0(X) \longrightarrow K_0'(X)$$

which associates to a vector bundle V on X its sheaf $\mathcal{O}(V)$. Also, if V is a vector bundle on X, then $\mathcal{F} \mapsto V \otimes \mathcal{F}$ is an exact functor from the category of coherent sheaves on X to itself, and one obtains a pairing

$$K_0(X) \otimes K_0'(X) \longrightarrow K_0'(X)$$

making $K_0'(X)$ a $K_0(X)$-module.

For non-singular algebraic varieties X, the natural homomorphism h, $h : K_0(X) \longrightarrow K_0'(X)$, which associates to a vector bundle V on X its sheaf $\mathcal{O}(V)$, is an isomorphism. This follows from the fact that for these varieties a coherent sheaf admits a finite resolution by algebraic vector bundles. Actually, we do not care too much about the words holomorphic and algebraic, because the whole construction is algebraic and for non-singular varieties one can use GAGA if necessary. In general, we will stick to the algebraic language.

Now let $f : X \longrightarrow Y$ be a proper morphism between the non-singular algebraic varieties X and Y, and let V be a vector bundle on X. Then

$$f_!([V]) = \sum_{q=0}^{n} (-1)^q [\mathrm{R}^q f_*(V)], \ (n = \dim(X)),$$

may be considered as an element of the Grothendieck group $K_0(Y)$. Thus one obtains a homomorphism

$$f_! : K_0(X) \longrightarrow K_0(Y).$$

Furthermore, the Chern character ch defines a homomorphism, also written ch,

$$\mathrm{ch} : K_0(X) \longrightarrow H^{2*}(X, \mathbf{Q}(*)).$$

(For the general construction of Chern characters, also in higher K-theory, we refer to Section 4.6.)

We will often write ch($\star$) for ch([$\star$]).

Grothendieck's Riemann-Roch Theorem reads as follows:

Theorem 4.1.2 (Grothendieck) *Let $f : X \longrightarrow Y$ be a proper morphism between non-singular algebraic varieties X and Y, and let $[V] \in K_0(X)$. Then:*

$$\mathrm{ch}(f_!([V])) \cup \mathrm{td}(Y) = f_*(\mathrm{ch}(V) \cup \mathrm{td}(X)).$$

Remark 4.1.2 Grothendieck's Riemann-Roch Theorem is an equality in the cohomology ring $H^{2*}(Y, \mathbf{Q}(\star))$, but it equally holds in other cohomology theories with a good notion of Chern classes etc. In particular, the theorem can be formulated for the Poincaré duality theories as discussed in the previous chapter.

Remark 4.1.3 The theorem can be interpreted by saying that the mapping from $K_0(X)$ to $H^\bullet(X)$ $(= H^{2*}(X, \mathbf{Q}(\star)))$ given by $[V] \mapsto \mathrm{ch}(V) \cup \mathrm{td}(X)$ is a natural transformation of covariant functors.

4.2 Adams Operations

For non-singular X, the operations of exterior powers of vector bundles endow $K_0(X)$ with extra structure, which makes $K_0(X)$ a so-called λ-ring. The general definition is:

Definition 4.2.1 *A unitary commutative ring R, endowed with a series of maps $\lambda^i : R \longrightarrow R$, $i \in \mathbf{N}$, is called a* λ-ring *if:*
(i) $\lambda^0(x) = 1$ *and* $\lambda^1(x) = x$, $\forall x \in R$.
(ii) $\lambda^k(x+y) = \sum_{i=0}^{k} \lambda^i(x)\,\lambda^{k-i}(y)$, $\forall x, y \in R$.

Introducing the formal power series

$$\lambda_t(x) = \sum_{i=0}^{\infty} \lambda^i(x)t^i \text{ and } \psi_{-t}(x) = -t\frac{d}{dt}(\log \lambda_t(x)),$$

one may define the so-called Adams operations $\psi^k : R \longrightarrow R$, $k \in \mathbf{N}$, by the formula

$$\psi_t(x) = \sum_{i=0}^{\infty} (-1)^i \psi^i(x) t^i .$$

One can verify that $\psi^1(x) = x$, $\forall x \in R$, and that the ψ^k's are group homomorphisms.

Example 4.2.1 $R = K_0(X)$ for a smooth scheme X over a field k. and $\mathcal{L} \longrightarrow X$ is a line bundle, defining a class $[\mathcal{L}] \in K_0(X)$. Then $\lambda_t([\mathcal{L}]) = 1 + [\mathcal{L}]t$ and $\psi^k([\mathcal{L}]) = [\mathcal{L}]^{\otimes k}$. The Adams operations

$$\psi^k : K_0(X) \longrightarrow K_0(X)$$

are actually ring (and even λ-ring) endomorphisms. Moreover, one has the identity $\psi^k \circ \psi^l = \psi^l \circ \psi^k = \psi^{kl}$.

This example will be the source of inspiration for much of the sequel of this chapter. There is already a lot of geometric theory hidden in it. More precisely, let $K_0(X)^{(n)}$ denote the subgroup of $K_0(X) \otimes \mathbf{Q}$, consisting of elements x such that $\psi^k(x) = k^n x$ for $\forall k \geq 1$. Beilinson defines:

Definition 4.2.2 *The* $(2n^{th})$ motivic cohomology *of* X *is*

$$H^{2n}_{\mathcal{M}}(X, \mathbf{Q}(n)) = \{x \in K_0(X) \otimes \mathbf{Q} | \psi^k(x) = k^n x, \forall k \geq 1\}.$$

A theorem of Grothendieck says:

Theorem 4.2.1 (Grothendieck) $H^{2n}_{\mathcal{M}}(X, \mathbf{Q}(n)) \cong CH^n(X) \otimes \mathbf{Q}$.

Taking Deligne-Beilinson cohomology $H^{2\bullet}_{\mathcal{D}}(X, \mathbf{Q}(\bullet))$ (or, with $\mathbf{R}$-coefficents, $H^{\bullet}_{\mathcal{D}}(X, \mathbf{R}(\bullet)))$ for X smooth, projective over the complex numbers $\mathbf{C}$, and writing $\mathrm{ch}_{n,2n}$ for the restriction of the Chern character $\mathrm{ch} \otimes \mathbf{Q}$ to $K_0(X)^{(n)}$, one obtains

$$\mathrm{ch}_{n,2n} : H^{2n}_{\mathcal{M}}(X, \mathbf{Q}(n)) \longrightarrow H^{2n}_{\mathcal{D}}(X, \mathbf{Q}(n)) \quad (\text{or } H^{2n}_{\mathcal{D}}(X, \mathbf{R}(n)))\,,$$

and it can be shown that this corresponds to the cycle map $\mathrm{cl}_{\mathcal{D}}$. Surjectivity of $\mathrm{ch}_{n,2n}$ would imply the truth of the famous Hodge Conjecture. We come back to this topic in the sequel, (cf. Chapter 8).

4.3 Riemann-Roch for Singular Varieties

For singular varieties P.Baum, W. Fulton and R. MacPherson [BFM] proved the following theorem, stressing the role of homology.

Theorem 4.3.1 *For a Poincaré duality theory* $(H^{\bullet}, H_{\bullet})$ *on the category of quasi-projective schemes (of finite type over a field k) there is a unique natural transformation*

$$\tau : K'_0 \longrightarrow H_{\bullet}\,,$$

such that:

(i) For any X the diagram

$$\begin{array}{ccc} K_0(X) \otimes K_0'(X) & \xrightarrow{\otimes} & K_0'(X) \\ \downarrow \mathrm{ch} \otimes \tau & & \downarrow \tau \\ H^\bullet(X) \otimes H_\bullet(X) & \xrightarrow{\cap} & H_\bullet(X) \end{array}$$

is commutative.

(ii) If X is non-singular and has structure sheaf $\mathcal{O}_X$, then

$$\tau([\mathcal{O}_X]) = \mathrm{td}(X) \cap \eta_X \,,$$

where η_X is the fundamental class of X in $H_\bullet$.

4.4 Higher Algebraic K-Theory

For any scheme X, Grothendieck's $K_0(X)$ and $K_0'(X)$ can be considered as the fundamental groups π_1 (with suitable base point) of the classifying spaces of some abstractly defined categories associated to the categories of algebraic vector bundles and of coherent sheaves on X, respectively. This fact inspired D. Quillen [Qu] to define higher algebraic K-groups $K_i(X)$ and $K_i'(X)$, $i = 1, 2, 3, \ldots$, as the higher homotopy groups π_{i+1} of the aforementioned classifying spaces.

In the sixties and early seventies various attempts by many mathematicians were made to define higher K-groups, and as one of the lasting results there is Quillen's definition of higher K-groups of rings R, the $K_i(R)$. These are also defined as homotopy groups and one proves that they are isomorphic with the higher K-groups of the affine scheme Spec(R), the $K_i(\mathrm{Spec}(R))$, alluded to in the first paragraph of this section. J-P. Jouanolou [Jo] showed that for suitable regular, quasi-projective schemes X there always exist a vector bundle $V \longrightarrow X$ and an affine V-torsor $p : W \longrightarrow X$ such that $K_*(X) \cong K_*(W)$.

Quillen's general theory gives a framework with some very useful properties that will be discussed in the following pages. Nonetheless, e.g. for the construction of a Chern character in higher algebraic K-theory, the higher K-groups of rings as defined before the invention of the general formalism are sometimes more appropriate or easier to handle. Therefore we give a brief sketch of both formalisms.

First the ring case. Let R be a ring and let $\mathrm{GL}_n(R)$ denote the group of invertible $n \times n$-matrices with entries in R. $\mathrm{GL}_n(R)$ contains the subgroup $\mathrm{E}_n(R)$ generated by the elementary matrices e_{ij}^{λ}, $\lambda \in R$, $1 \leq i \neq j \leq n$, defined by

$$(e_{ij}^{\lambda})_{kl} = \begin{cases} 1 & \text{if } k = l \\ \lambda & \text{if } k = i \text{ and } l = j \\ 0 & \text{otherwise.} \end{cases}$$

For $n \geq 3$ the group $\mathrm{E}_n(R)$ is perfect, i.e. equal to its commutator subgroup. Denote by $\mathrm{GL}(R) = \varinjlim \mathrm{GL}_n(R)$, where $\mathrm{GL}_n(R) \longrightarrow \mathrm{GL}_{n+1}(R)$ embeds $\mathrm{GL}_n(R)$ into the upper left part of $\mathrm{GL}_{n+1}(R)$. Similarly we will write $\mathrm{E}(R) = \varinjlim \mathrm{E}_n(R)$. Then $\mathrm{E}(R)$ is normal in $\mathrm{GL}(R)$.

Let $\mathrm{BGL}_n(R)$ $(n \geq 3)$ and $\mathrm{BGL}(R)$ be the classifying spaces, then $\mathrm{BGL}(R) = \varinjlim \mathrm{BGL}_n(R)$. From now on we will consider only $\mathrm{BGL}(R)$, $\mathrm{E}(R)$, ... etc., the corresponding statements for $\mathrm{BGL}_n(R)$, $\mathrm{E}_n(R)$, ... etc. being obvious.

One may notice that $\mathrm{BGL}(R)$ is an Eilenberg-MacLane space of type $K(\mathrm{GL}(R), 1)$. To this space (with base point $\star$) we will associate in a functorial and universal way, up to homotopy, a space $\mathrm{BGL}(R)^+$ which is connected and has fundamental group

$$\pi_1(\mathrm{BGL}(R)^+, \star) \cong \mathrm{GL}(R)/\mathrm{E}(R),$$

such that the map $\pi_1(\mathrm{BGL}(R), \star) \longrightarrow \pi_1(\mathrm{BGL}(R)^+, \star)$ corresponds to $\mathrm{GL}(R) \longrightarrow \mathrm{GL}(R)/\mathrm{E}(R)$, and the map $\imath : \mathrm{BGL}(R) \longrightarrow \mathrm{BGL}(R)^+$ induces an isomorphism on homology (with values in a local coefficient system $\mathcal{L}$):

$$H_*(\mathrm{BGL}(R), \imath^*(\mathcal{L})) \xrightarrow{\sim} H_*(\mathrm{BGL}(R)^+, \mathcal{L}).$$

The construction of $\mathrm{BGL}(R)^+$ is called Quillen's $+$-construction.

It goes as follows: Write X for $\mathrm{BGL}(R)$, N for $\mathrm{E}(R)$ and π for $\pi_1(X, \star) \cong \mathrm{GL}(R)$, and with these shorthand notations, let A be any subset of N such that the smallest normal subgroup of π containing A, is equal to N. An element $\alpha \in N$ defines a continuous map $a_\alpha : \mathrm{S}^1 \longrightarrow X$

with homotopy class $\alpha \in \pi$. Using these a_α's one can attach 2-cells e^2_α to the cell complex X to obtain the complex $X_1 = X \cup_{\alpha\in A} e^2_\alpha$. Van Kampen's theorem says that $\pi_1(X_1) = \pi_1(X)/N = \pi/N$ and the inclusion $\imath_1 : X \hookrightarrow X_1$ induces a surjection $\pi \longrightarrow \pi/N$. One has the cartesian diagram

$$\begin{array}{ccc} \hat{X} & \overset{\imath}{\hookrightarrow} & \tilde{X}_1 \\ \downarrow & \square & \downarrow \\ X & \overset{\imath_1}{\hookrightarrow} & X_1 \end{array}$$

where $\tilde{X}_1$ denotes the universal covering of X_1, and $\hat{X}$ is the subspace of $\tilde{X}_1$ lying over X. Both vertical arrows have the same fibre $\pi_1(X_1) = \pi/N$. The exact homotopy sequence for the left fibration gives

$$0 \longrightarrow \pi_1(\hat{X}) \longrightarrow \pi_1(X) \longrightarrow \pi/N \longrightarrow 0,$$

thus $\pi_1(\hat{X}) \xrightarrow{\sim} N$. There is the commutative diagram for $(\tilde{X}_1, \hat{X})$:

$$\begin{array}{ccccc} \pi_2(\tilde{X}_1) & \longrightarrow & \pi_2(\tilde{X}_1, \hat{X}) & \longrightarrow & \pi_1(\hat{X}) = N \\ \rho\downarrow & & \downarrow & & \downarrow \\ H_2(\tilde{X}_1, \mathbf{Z}) & \xrightarrow{j} & H_2(\tilde{X}_1, \hat{X}) & \longrightarrow & H_1(\hat{X}, \mathbf{Z}). \end{array}$$

N is perfect, thus $H_1(\hat{X}, \mathbf{Z}) = \pi_1(\hat{X})/[\pi_1(\hat{X}), \pi_1(\hat{X})] = N/[N,N] = 0$. So, j is surjective. Also, because $\tilde{X}_1$ is simply connected, the Hurewicz homomorphism ρ is an isomorphism. $H_2(X_1, X)$ is the free abelian group generated by the 2-cells attached to X to get X_1. Therefore $H_2(\tilde{X}_1, \hat{X})$ is a free $\mathbf{Z}[\pi/N]$-module with basis $\{e^2_\alpha\}_{\alpha\in A}$. Using

$$\pi_2(\tilde{X}_1) \xrightarrow{\sim} H_2(\tilde{X}_1, \mathbf{Z}) \xrightarrow{\text{surj.}} H_2(\tilde{X}_1, \hat{X}),$$

one can find continuous maps $b_\alpha : S^2 \longrightarrow X_1$, $\alpha \in A$, such that $\{j \circ \rho[b_\alpha]\} = \{e^2_\alpha\}$. Now we can attach 3-cells to X_1 to obtain

$$X^+ = X_1 \cup_{\alpha\in A} e^3_\alpha .$$

It is then clear that $\imath : X \hookrightarrow X^+$ induces the projection $\pi \longrightarrow \pi/N$. One can prove that for a local coefficient system $\mathcal{L}$, defined by a $\mathbf{Z}[\pi/N]$-module, the relative homology $H_*(X^+, X; \mathcal{L})$ vanishes, so X and X^+ have the same homology. For details we refer to [Lo].

Definition 4.4.1 $K_i(R) = \pi_i(\mathrm{BGL}(R)^+)$, $i \geq 1$.

Thus the K_i, $i \geq 1$, define covariant functors from $\mathcal{A}nn$, the category of rings, to $\mathcal{A}b$, the category of abelian groups.

Example 4.4.1 $K_1(R) = \mathrm{GL}(R)/\mathrm{E}(R)$, the Whitehead group.

Example 4.4.2 $K_2(R) = H_2(\mathrm{E}(R), \mathbf{Z})$, Milnor's K_2.

An interesting account of K_0, K_1 and K_2 for rings is [Mi].

We close with a few results on the K-theory of fields k.

Proposition 4.4.1 *$K_0(k) = \mathbf{Z}$ and $K_1(k) = k^*$.*

Theorem 4.4.1 (Matsumoto) *$K_2(k)$ has a presentation in terms of generators and relations as follows:*
The generators are symbols $\{x, y\}$, for x and y in k^*, subject to the following relations and their consequences:
(i) $\{x, 1-x\} = 1$ *for* $x \neq 0, 1$.
(ii) $\{x_1x_2, y\} = \{x_1, y\}\{x_2, y\}$.
(iii) $\{x, y_1y_2\} = \{x, y_1\}\{x, y_2\}$.

Theorem 4.4.2 (Tate) *$K_2(\mathbf{Q})$ is canonically isomorphic with the direct sum $A_2 \oplus A_3 \oplus A_5 \oplus \ldots$, where A_2 is the cyclic group $\{\pm 1\}$ and where $A_p = (\mathbf{Z}/p\mathbf{Z})^*$ for the odd primes p.*

Theorem 4.4.3 (Tate) *Let K be a number field with ring of integers $\mathcal{O}_K$. Then $K_2(K)$ is generated by those symbols $\{x, y\}$ for which x and y are relatively prime elements of $\mathcal{O}_K$.*

For later use we state the following result, due to Quillen.

Theorem 4.4.4 (Quillen) *Let K be a number field with ring of integers $\mathcal{O}_K$. Then $K_i(\mathcal{O}_K)$ is finitely generated and*

$$K_i(\mathcal{O}_K) \otimes \mathbf{Q} = K_i(K) \otimes \mathbf{Q} \quad \text{for} \quad i \geq 1.$$

Next we consider Quillen's general formalism and some basic properties of higher algebraic K-groups. To this end, let $\mathcal{C}$ be an additive category that is embedded as a full subcategory of an abelian category $\mathcal{A}$, and such that if A in $\mathcal{A}$ has a subobject A' with A' and A/A' isomorphic with objects of $\mathcal{C}$, then A itself is isomorphic with an object of $\mathcal{C}$. Also, let $\mathcal{E}$ be the class of sequences

$$0 \longrightarrow A' \xrightarrow{i} A \xrightarrow{j} A'' \longrightarrow 0$$

in $\mathcal{C}$ which are exact in $\mathcal{A}$. A morphism in $\mathcal{C}$ is called an **admissible monomorphism**, resp. **admissible epimorphism** if it occurs as the i, resp. j of some sequence in $\mathcal{E}$. $\mathcal{E}$ enjoys the following properties:

(a) Any sequence in $\mathcal{C}$ isomorphic with a sequence in $\mathcal{E}$ belongs to $\mathcal{E}$. For any A', A'' in $\mathcal{C}$, the sequence

$$0 \longrightarrow A' \xrightarrow{(id,o)} A' \oplus A'' \xrightarrow{pr_2} A'' \longrightarrow 0$$

belongs to $\mathcal{E}$. For any sequence in $\mathcal{E}$, i is a kernel for j and j is a cokernel for i in $\mathcal{C}$.

(b) The class of admissible epimorphisms is closed under composition and under base change by arbitrary morphisms in $\mathcal{C}$. Dually, one has an analogous statement for admissible monomorphisms.

(c) Let $A \longrightarrow A''$ be a morphism in $\mathcal{C}$ possessing a kernel. If there is a morphism $B \longrightarrow A$ in $\mathcal{C}$ such that $B \longrightarrow A \longrightarrow A''$ is an admissible epimorphism, then $A \longrightarrow A''$ is an admissible epimorphism. Dually for admissible monomorphisms.

Definition 4.4.2 *An additive category $\mathcal{C}$ equipped with a class $\mathcal{E}$ as above, called the class of* **(short) exact sequences** *in $\mathcal{C}$, such that (a), (b) and (c) hold, is called an* **exact category**. *An* **exact functor** *between two exact categories is an additive functor carrying exact sequences into exact sequences.*

Example 4.4.3 An abelian category is an exact category. The pseudo-abelian (or Karoubian) envelope (cf. Chapter 8) of an additive category is exact.

Example 4.4.4 The category $\mathcal{V}ectb(X)$ of vector bundles on a scheme X is exact.

Example 4.4.5 The category $\mathcal{C}oh(X)$ of coherent sheaves on a noetherian, separated scheme X is exact.

To an exact category $\mathcal{C}$ one associates a new category $Q\mathcal{C}$ whose objects are the same as those of $\mathcal{C}$, but whose morphisms are isomorphism classes of diagrams $A \xleftarrow{j} B \xrightarrow{i} A'$, where j is an admissible epimorphism and i is an admissible monomorphism. Isomorphisms should induce the identity on A and A'. Composition of two morphisms in $Q\mathcal{C}$, say $A \xleftarrow{j} B \xrightarrow{i} A'$ and $A' \xleftarrow{j'} C \xrightarrow{i'} A''$ gives the morphism $A \longleftarrow B \times_{A'} C \longrightarrow A''$, represented by $j \circ pr_1$ and $i' \circ pr_2$ as follows:

$$\begin{array}{ccccc}
B \times_{A'} C & \xrightarrow{pr_2} & C & \xrightarrow{i'} & A'' \\
{\scriptstyle pr_1}\downarrow & & \downarrow{\scriptstyle j'} & & \\
B & \longrightarrow & A' & & \\
{\scriptstyle j}\downarrow & & & & \\
A & & & &
\end{array}$$

This may be compared with the composition of morphisms in derived categories.

Assume from now on that the category $\mathcal{C}$ is small and exact. Then $Q\mathcal{C}$ is small too, and its **nerve**, denoted $NQ\mathcal{C}$ is the semi-simplicial set whose p-simplices $X_p^\bullet$ are diagrams in $Q\mathcal{C}$ of the form

$$X_p^\bullet = (X_0 \longrightarrow X_1 \longrightarrow \dots \longrightarrow X_p)\,.$$

The ith face $\partial_i(X_p^\bullet)$ is defined as

$$\partial_i(X_p^\bullet) = (X_0 \longrightarrow \dots \longrightarrow X_{i-1} \longrightarrow X_{i+1} \longrightarrow \dots \longrightarrow X_p)\,,$$

and the ith degeneracy $\sigma_i(X_p^\bullet)$ is given by

$$\sigma_i(X_p^\bullet) = (X_0 \longrightarrow \dots \longrightarrow X_i \xrightarrow{\text{id}} X_i \longrightarrow \dots \longrightarrow X_p)\,,$$

$i = 0, 1, \dots, p$. Here by $X_k \longrightarrow X_l$ is meant, of course, a diagram of the form $X_k \xleftarrow{j} Y \xrightarrow{i} X_l$.

Next we describe how one can associate a topological space B$\mathcal{QC}$ to the complex N$\mathcal{QC}$. To see how this is done, let N$\mathcal{QC}_n$ denote the collection of n-simplices of N$\mathcal{QC}$, $n = 0, 1, \ldots$. Furthermore, Δ^n will denote the topological n-simplex in $\mathbf{R}^{n+1}$, defined by

$$\Delta^n = \Big\{(t_0, t_1, \ldots, t_n) | 0 \le t_i \le 1, \sum_i t_i = 1\Big\}.$$

Define maps $\delta_i : \Delta^{n-1} \longrightarrow \Delta^n$ and $s_i : \Delta^{n+1} \longrightarrow \Delta^n$ by

$$\begin{aligned} \delta_i(t_0, \ldots, t_{n-1}) &= (t_0, \ldots, t_{i-1}, 0, t_i, \ldots, t_{n-1}), \\ s_i(t_0, \ldots, t_{n+1}) &= (t_0, \ldots, t_i + t_{i+1}, \ldots, t_{n+1}). \end{aligned}$$

Now endow the complex N$\mathcal{QC}$ with the discrete topology and form the disjoint union

$$\overline{\mathrm{N}\mathcal{QC}} = \bigcup_{n \ge 0} (\mathrm{N}\mathcal{QC}_n \times \Delta^n).$$

On $\overline{\mathrm{N}\mathcal{QC}}$ one defines an equivalence relation $\sim$ by

$$\begin{aligned} (\partial_i(X_n^\bullet), y_{n-1}) &\sim (X_n^\bullet, \delta_i y_{n-1}), \quad X_n^\bullet \in \mathrm{N}\mathcal{QC}_n, \ y_{n-1} \in \Delta^{n-1}, \text{ and} \\ (\sigma_i(X_n^\bullet), y_{n+1}) &\sim (X_n^\bullet, s_i y_{n+1}), \quad X_n^\bullet \in \mathrm{N}\mathcal{QC}_n, \ y_{n+1} \in \Delta^{n+1}. \end{aligned}$$

The identification space $\overline{\mathrm{N}\mathcal{QC}}/\!\sim$ is called the **geometric realization** of N$\mathcal{QC}$. This construction can be carried out for any simplicial set (complex). Thus it defines a covariant functor

$$T : \mathcal{K} \longrightarrow \mathcal{T}op,$$

where $\mathcal{K}$, resp. $\mathcal{T}op$, is the category of complexes, resp. of topological spaces (with obvious morphisms). One can show that this realization functor T is actually adjoint to the total singular complex functor $S : \mathcal{T}op \longrightarrow \mathcal{K}$. Furthermore, for a complex $K \in \mathcal{O}b(\mathcal{K})$ and a topological space $X \in \mathcal{O}b(\mathcal{T}op)$ there are one-to-one correspondences between homotopy classes of simplicial maps $K \longrightarrow S(X)$ and of continuous maps $T(K) \longrightarrow X$.

The geometric realization of N$\mathcal{QC}$ is called the **classifying space** of $\mathcal{QC}$ and is written B$\mathcal{QC}$. For an object X of $\mathcal{QC}$, X may be used to denote the corresponding 0-cell of B$\mathcal{QC}$, and one can define the **homotopy groups** $\pi_i(\mathrm{B}\mathcal{QC}, X)$, $i \ge 1$, of $\mathcal{QC}$ with base point X. It is costumary to write $\pi_0(\mathrm{B}Q\mathcal{C}, X)$ for the set of components of $Q\mathcal{C}$, pointed by the component that contains X.

Let 0 be a zero-object of $\mathcal{C}$. Then, also, 0 is an object of $Q\mathcal{C}$, and we arrive at Quillen's definition of the higher K-groups of the category $\mathcal{C}$.

Definition 4.4.3 (Quillen) *Let $\mathcal{C}$ be an exact category with classifying space* $\mathrm{B}Q\mathcal{C}$ *and base point* 0. *Then*

$$K_i(\mathcal{C}) = \pi_{i+1}(\mathrm{B}Q\mathcal{C}, 0)\,,\ i = 0, 1, \ldots .$$

Example 4.4.6 For a unitary ring R let $\mathcal{P}(R)$ denote the category of finitely generated projective (left) R-modules. This is an additive category that becomes exact when one takes as exact sequences those sequences that are exact in the category of all R-modules, which is abelian. Then, define $K_i(R) = K_i(\mathcal{P}(R))$. One can prove that this definition coincides with the one coming from the +-construction. Also, because of the equivalence $\mathcal{P}(R) \simeq \mathcal{V}ectb(\mathrm{Spec}(R))$, where $\mathcal{V}ectb$ denotes the category of vector bundles, one has

$$K_i(\mathrm{Spec}(R)) = K_i(R)\,,\ i = 0, 1, \ldots .$$

As we are mainly interested in algebro-geometric aspects of K-theory, we have to consider the category $\mathcal{S}ch$ of noetherian, separated schemes X and the corresponding K-groups. More precisely, one has the following

Definition 4.4.4 *Let X be a noetherian, separated scheme. Then*

$$K_i(X) = K_i(\mathcal{V}ectb(X)) \text{ and } K_i' = K_i(\mathcal{C}oh(X))$$

where $\mathcal{V}ectb(X)$, resp. $\mathcal{C}oh(X)$ denote the exact categories of vector bundles, resp. coherent sheaves on X.

The functors K_i, $K_i' : \mathcal{S}ch \longrightarrow \mathcal{A}b$ have the following properties:

(1) When X is regular, $K_i(X) \cong K_i'(X)$, $\forall i \geq 0$.

(2) An arbitrary morphism $f : X \longrightarrow Y$ gives rise to a homomorphism $f^* : K_i(Y) \longrightarrow K_i(X)$, and whenever f is flat, f induces a homomorphism $f^* : K_i'(Y) \longrightarrow K_i'(X)$, $\forall i \geq 0$.

(3) For a filtered projective system $\{X_j\}_{j\in J}$ of schemes with affine transition morphisms, and with $X = \varprojlim X_j$, one has $K_i(X) = \varinjlim K_i(X_j)$. Whenever, in addition, the transition morphisms are flat, then $K_i'(X) = \varinjlim K_i'(X_j)$.

(4) For proper morphisms $f : X \longrightarrow Y$, one obtains homomorphisms $f_* : K_i'(X) \longrightarrow K_i'(Y)$, $\forall i \geq 0$.

(5) There are pairings $K_0(X) \otimes K_i'(X) \longrightarrow K_i'(X)$ giving $K_i'(X)$ the structure of a $K_0(X)$-module.

(6) (Projection Formula) Let $f : X \longrightarrow Y$ be proper and flat. Then for any $x \in K_0(X)$ and $y \in K_i'(X)$ one has the formula

$$f_*(x \cdot f^*(y)) = f_!(x) \cdot y$$

in $K_i'(Y)$, where we have written $f_!(x)$ for the image of the homomorphism $f_! : K_0(X) \longrightarrow K_0(Y)$.

(7) Let $\imath : Z \hookrightarrow X$ be a closed immersion and let $j : U \hookrightarrow X$ be the open immersion of the complement U of Z in X. Then there is an exact sequence

$$\ldots \longrightarrow K_{i+1}'(U) \longrightarrow K_i'(Z) \xrightarrow{\imath_*} K_i'(X) \xrightarrow{j^*} K_i'(U) \longrightarrow \ldots .$$

(8) (Homotopy Property) Let $f : X \longrightarrow Y$ be a flat map with affine fibres, e.g. a vector bundle or a torsor under a vector bundle. Then $f^* : K_i'(Y) \longrightarrow K_i'(X)$ is an isomorphism for all $i \geq 0$.

(9) Let V be a rank r vector bundle on X with associated projective bundle $f : \mathbf{P}(V) \longrightarrow X$. Then one will have a $K_0(\mathbf{P}(V))$-module isomorphism $\phi : K_0(\mathbf{P}(V)) \otimes_{K_0(X)} K_i'(X) \longrightarrow K_i'(\mathbf{P}(V))$ given by

$$\phi(v \otimes x) = v.f^*(x) .$$

Notwithstanding the basic importance of the above properties, the next result plays such an essential role on several places in the sequel that we formulate it as a theorem.

Theorem 4.4.5 (Quillen) *Let $X^{(p)}$ denote the set of points of codimension p in X. There is a spectral sequence*

$$E_1^{p,q}(X) = \coprod_{x \in X^{(p)}} K_{-p-q}(k(x)) \Longrightarrow K'_{-p-q}(X) ,$$

which is convergent when X has finite Krull dimension. Moreover, this spectral sequence is contravariant for flat morphisms. For a projective limit $X = \varprojlim X_j$ *with affine flat transition morphisms, the spectral sequence for X is the inductive limit of the spectral sequences of the* X_j.

One can also sheafify the K-groups to obtain the sheaves $\mathcal{K}_i$. A fundamental result relating these sheaves to the Chow group is given by the following theorem, which for $i = 2$ is due to S. Bloch and in general to D. Quillen.

Theorem 4.4.6 (Bloch-Quillen) *Let X be a regular scheme of finite type over a field. Then* $H^i(X, \mathcal{K}_i) \cong CH^i(X)$.

For a regular, noetherian scheme X of finite Krull dimension, and a closed immersion $Y \hookrightarrow X$, one can define the K-groups of X with support in Y by:

$$K_i^Y(X) = \pi_{i+1}(fibre(\mathrm{BQ}\mathcal{V}ectb(X) \longrightarrow \mathrm{BQ}\mathcal{V}ectb(X\backslash Y)),$$

where the fibre means the homotopy fibre of the fibration

$$\mathrm{BQ}\mathcal{V}ectb(X) \longrightarrow \mathrm{BQ}\mathcal{V}ectb(X\backslash Y).$$

In particular, one has $K_i^X(X) = K_i(X)$. For smooth schemes the isomorphism $K_i(X) \xrightarrow{\sim} K_i'(X)$ generalizes to:

Theorem 4.4.7 (Purity Theorem) *If X is smooth and $Y \hookrightarrow X$ is a closed subscheme, then there is a canonical isomorphism*

$$K_i^Y(X) \xrightarrow{\sim} K_i'(Y).$$

In particular, if Y is regular this means that $K_i(Y) \simeq K_i^Y(X)$.

4.5 Adams Operations in Higher Algebraic K-Theory

Using tensor products of matrices in $\mathrm{GL}_p(R)$ and $\mathrm{GL}_q(R)$, where R is a commutative unitary ring, and taking limits gives a map

$$\hat{\otimes} : \mathrm{BGL}(R)^+ \times \mathrm{BGL}(R)^+ \longrightarrow \mathrm{BGL}(R)^+$$

which, via the homeomorphism $S^{n+m} \simeq S^n \wedge S^m$, induces bilinear products

$$K_i(R) \times K_j(R) \xrightarrow{\cup} K_{i+j}(R).$$

Sheafifying in a suitable way, this gives products in general K-theory

$$K_i^Y(X) \times K_j^Y(X) \longrightarrow K_{i+j}^Y(X).$$

One also has an augmentation map $\varepsilon : K_i^Y(X) \longrightarrow H_Y^0(X, \mathbf{Z})$ which is zero if $i > 0$. One can define operations

$$\lambda^k : K_i^Y(X) \longrightarrow K_i^Y(X),$$

which make the $K_i^Y(X)$, $i \geq 0$, so-called $K_0^Y(X)$-λ-algebras. This implies that one can introduce new operations, γ^k, $k \geq 0$, by the formula

$$\gamma_t(x) = 1 + \sum_{i=1}^{\infty} \gamma^i(x)t^i = \lambda_{t/(1-t)}(x) = 1 + \sum_{i=1}^{\infty} \lambda^i(x)t^i/(1-t)^i,$$

and also Adams operations ψ^k, $k \geq 0$, by the recurrence

$$\sum_{i=0}^{k-1} (-1)^i \psi^{k-i}\lambda^i + (-1)^k k\lambda^k = 0.$$

For the product structure $K_i^Y(X) \times K_j^Y(X) \longrightarrow K_{i+j}^Y(X)$ the ψ^k and the γ^k enjoy the properties

$$\psi^k(xy) = \psi^k(x)\psi^k(y), \text{ for } k > 1 \text{ and} \tag{4.1}$$

$$\gamma^k(xy) = \sum_{k'+k''=k} -\frac{(k-1)!}{(k'-1)!(k''-1)!} \gamma^{k'}(x)\gamma^{k''}(y), \text{ if } i.j \neq 0. \tag{4.2}$$

Write $F_\gamma^0 K_i^Y(X) = K_i^Y(X)$ and let $F_\gamma^m K_i^Y(X)$ denote the subgroup of $K_i^Y(X)$ generated by the products $\gamma^{i_1}(x_1)\gamma^{i_2}(x_2)\ldots\gamma^{i_l}(x_l)$, where $\varepsilon(x_1) = \ldots = \varepsilon(x_l) = 0$ and $i_1 + \cdots i_l \geq m$. The corresponding grading is denoted by

$$Gr_\gamma^m K_i^Y(X) = F_\gamma^m K_i^Y(X)/F_\gamma^{m+1} K_i^Y(X). \tag{4.3}$$

Remark 4.5.1 Actually, to have a better defined γ-filtration and corresponding grading, one should neglect the torsion and take the γ-filtration of $K_i^Y(X) \otimes \mathbf{Q}$. Then the same filtration arises by taking $F_\gamma^m K_i^Y(X) \otimes \mathbf{Q}$ as the subgroup generated by the elements $\gamma^n(x)$, $n \geq m$. In the sequel we will assume that the torsion has been eliminated in this way.

Remark 4.5.2 A useful observation, frequently quoted in what follows, is that the action of the Adams operation ψ^k on $Gr_\gamma^m K_i^Y(X) \otimes \mathbf{Q}$ is just multiplication by k^m.

Let S be a regular, irreducible, noetherian scheme of finite Krull dimension. Denote by $\mathcal{V}_S$ the category of quasi-projective schemes over S. Let $X \in \mathcal{O}b(\mathcal{V}_S)$ and let $\imath : X \hookrightarrow M$ be a closed immersion of X in the smooth scheme M over S. Then $\imath^* : K_i'(X) \xrightarrow{\sim} K_i^X(M)$. Define the filtration $F_\bullet$ of $K_i'(X) \otimes \mathbf{Q}$ by

$$F_m = F_m K_i'(X) \otimes \mathbf{Q} = \imath_*^{-1}(F_\gamma^{\dim(M)-m} K_i^X(M) \otimes \mathbf{Q}),$$

and write $Gr_m = F_m / F_{m+1}$ for the corresponding grading.

Definition 4.5.1 *Let $X \in \mathcal{O}b(\mathcal{V}_S)$. Then, for $i, j, n, p \in \mathbf{Z}$ one defines*

$$H_n(X, j) = Gr_j K_{n-2j}'(X) \otimes \mathbf{Q},$$

and for a closed immersion $Y \hookrightarrow X$ in $\mathcal{V}_S$, where X is smooth over S,

$$H_Y^p(X, i) = Gr_\gamma^i K_{2i-p}^Y(X) \otimes \mathbf{Q}.$$

The main result of this section is the following theorem, of fundamental importance in Beilinson's philosophy and due to C. Soulé [So]:

Theorem 4.5.1 (Soulé) *$(H^\bullet, H_\bullet)$, with the abbreviated notation $H^\bullet = H_Y^\bullet(X, \star)$ and $H_\bullet = H_\bullet(X, \star)$, is a Poincaré duality theory with supports.*

In this Poincaré duality theory the cap product is induced by the obvious pairing $\mathcal{V}ectb(X) \otimes \mathcal{C}oh(X) \longrightarrow \mathcal{C}oh(X)$.

4.6 Chern Classes in Higher Algebraic K-Theory

For any Poincaré duality theory $(H^\bullet, H_\bullet)$ there is a good theory of Chern classes, taking values in the coefficients of the $H^\bullet$ and $H_\bullet$. In what follows we shall often write $H^\bullet(X,\star)$, ... etc. for specific cohomology spaces, ... etc. of an underlying Poincaré duality theory.

The aim of this section is to recall the construction of Chern classes in higher algebraic K-theory. This will lead, in the next section, to H. Gillet's formulation of the Riemann-Roch Theorem for higher algebraic K-theory.

The construction of Chern classes in higher algebraic K-theory is realized in several steps, one of the most fundamental being the construction of natural homomorphisms

$$c_{i,j} : K_i(R) \longrightarrow H^{2j-i}(\mathrm{Spec}(R),\star),$$

associated to a family of classes $c_j^{(n)} \in H^{2j}(\mathrm{B}_\bullet\mathrm{GL}_n,\star)$. Here $\mathrm{B}_\bullet\mathrm{GL}_n$ denotes the classifying simplicial scheme

$$\mathrm{B}_\bullet\mathrm{GL}_n \ : \ \mathrm{Spec}(S) \leftleftarrows \mathrm{GL}_n \ \underset{\xleftarrow{id\otimes 1}}{\overset{\xleftarrow{id\otimes 1}}{\xleftarrow{\ \mu\ }}} \ \mathrm{GL}_n \times \mathrm{GL}_n \ \begin{smallmatrix}\longleftarrow\\ \longleftarrow\\ \longleftarrow\\ \longleftarrow\end{smallmatrix} \ \cdots,$$

where μ denotes multiplication for the group scheme GL_n.

Let $\mathbf{Z}[\mathcal{B}_\bullet\mathcal{GL}_n]$ denote the sheaf complex of integral group rings associated to $\mathrm{B}_\bullet\mathrm{GL}_n$. A stability result says that in degree $\geq -\frac{n-1}{2}$ the complexes $\mathbf{Z}[\mathcal{B}_\bullet\mathcal{GL}_n]$ and $\mathbf{Z}[\mathcal{B}_\bullet\mathcal{GL}]$ are quasi-isomorphic. If we assume that $\star$ is represented by some complex of (injective) sheaves on $\mathcal{V}_S$, then there is a general result which says that there is a natural identification

$$H^*(\mathrm{B}_\bullet\mathrm{GL}_n,\star) = \mathrm{Hom}_{\mathrm{D}\mathcal{V}_S}(\mathbf{Z}[\mathcal{B}_\bullet\mathcal{GL}_n],\star[*]),$$

where the Hom is taken in the derived category $\mathrm{D}\mathcal{V}_S$ of $\mathcal{V}_S$. This implies in particular that the classes $c_j^{(n)}$, which are assumed to be compatible with respect to the embeddings $\mathrm{GL}_n \longrightarrow \mathrm{GL}_{n+1}$, induce a morphism

$$c_j : \mathbf{Z}[\mathcal{B}_\bullet\mathcal{GL}] \longrightarrow \star[2j]$$

in the derived category $D\mathcal{V}_S$. Passing to global sections over the affine scheme Spec(R) then gives a homomorphism of complexes, again denoted by c_j,

$$c_j : \mathbf{Z}[\mathrm{B_\bullet GL}(R)] \longrightarrow \Gamma(\mathrm{Spec}(R), \mathcal{H}om(\mathbf{Z}[\mathcal{B}_\bullet \mathcal{GL}], \star[2j])),$$

and finally, one obtains homomorphisms

$$c_{i,j} : H_i(\mathrm{GL}(R), \mathbf{Z}) \longrightarrow H^{2j-i}(\mathrm{Spec}(R), \star).$$

Combining with the Hurewicz map

$$\rho_i : \pi_i(\mathrm{BGL}(R)^+) \longrightarrow H_i(\mathrm{BGL}(R)^+, \mathbf{Z})$$

we get, for $i \geq 1$, natural homomorphisms

$$c_{i,j} : K_i(R) \longrightarrow H^{2j-i}(\mathrm{Spec}(R), \star)$$

as the compositions

$$K_i(R) = \pi_i(\mathrm{BGL}(R)^+) \xrightarrow{\rho_i} H_i(\mathrm{BGL}(R)^+, \mathbf{Z}) =$$
$$= H_i(\mathrm{BGL}(R), \mathbf{Z}) = H_i(\mathrm{GL}(R), \mathbf{Z}) \xrightarrow{c_{i,j}} H^{2j-i}(\mathrm{Spec}(R), \star).$$

To extend this construction to arbitrary schemes X of $\mathcal{V}_S$ one uses several properties of the Poincaré duality theory.

First, there is the definition of the Chern classes of a rank r vector bundle V on X (or on a simplicial object $X_\bullet$ of $\mathcal{V}_S$). These are uniquely determined classes $c_j(V) \in H^{2j}(X, \star)$ with $c_0 = 1$ and $c_j = 0$ for $j > r$. The following proposition (cf. [Sc]) resumes some important properties of Chern classes.

Proposition 4.6.1 *Let V be a rank r vector bundle on the scheme X (or $X_\bullet$). Then there is a natural homomorphism*

$$c : K_0(X) \longrightarrow H^0(X, \mathbf{Z}) \times \left\{ (x_j) \in \prod_{j \geq 0} H^{2j}(X, \star) \,|\, x_0 = 1 \right\}$$

given by $c(V) = (r, c_0(V), c_1(V), \ldots)$. (Similarly for $X_\bullet$).
Moreover, the family of these homomorphisms for all X (or $X_\bullet$) is uniquely characterized by the fact that, for line bundles $\mathcal{L}$, one has

$$c([\mathcal{L}]) = (1, 1, c_1(\mathcal{L}), \ldots).$$

In particular, one obtains natural maps

$$c_{0,j} : K_0(X) \longrightarrow H^{2j}(X,\star),$$

defined by $c_{0,j}([V]) = c_j(V)$, $j \geq 0$.

Next, we take a special family of classes $c_j^{(n)} \in H^{2j}(\mathrm{B_\bullet GL}_n,\star)$ to assure universality, i.e. indepedence of the particular Poincaré duality theory $(H^\bullet, H_\bullet)$. To this end, let $U_\bullet^n$, resp. τ^n, be the universal, resp. trivial, rank n vector bundles on $\mathrm{B_\bullet GL}_n$, and let

$$u_n = [U_\bullet^n] - [\tau^n] \in K_0(\mathrm{B_\bullet GL}_n).$$

Then $u = \{u_n\}_n$ is an element of $\varprojlim K_0(\mathrm{B_\bullet GL}_n)$.

Definition 4.6.1 *The* universal Chern classes $c_j^{(n)} \in H^{2j}(\mathrm{B_\bullet GL}_n,\star)$ *are defied by*

$$c(u) = \{(0, c_0^{(n)}, c_1^{(n)}, \ldots)\}_n.$$

Using Jouanolou's result, which says that for a quasi-projective scheme X there exists a vector bundle V on X and an affine V-torsor inducing an isomorphisms of K-groups, one can now construct, for any scheme $X \in \mathcal{O}b(\mathcal{V}_S)$, universal Chern class maps $c_{i,j}$ just as above.

Conclusion 4.6.1 *For $X \in \mathcal{O}b(\mathcal{V}_S)$ one has (uniquely determined) Chern class maps*

$$c_{i,j} : K_i(X) \longrightarrow H^{2j-i}(X,j).$$

With respect to the Adams operations and products on the higher K-groups, one has

Proposition 4.6.2 *The following diagrams commute*

$$\begin{array}{ccc}
K_i(X) & \xrightarrow{\psi^k} & K_i(X) \\
\downarrow{\scriptstyle c_{i,j}} & & \downarrow{\scriptstyle c_{i,j}} \\
H^{2j-i}(X,j) & \xrightarrow{.k^j} & H^{2j-i}(X,j)
\end{array}$$

$$\begin{array}{ccc} K_i(X)\otimes K_m(X) & \longrightarrow & K_{i+m}(X) \\ {\scriptstyle \sum_{j+n=k} c_{i,j}\otimes c_{m,n}}\downarrow & & \downarrow{\scriptstyle c_{i+m,k}} \\ \bigoplus_{j+n=k} H^{2j-i}(X,j)\otimes H^{2n-m}(X,n) & \xrightarrow{\cup_k} & H^{2k-(i+m)}(X,k) \end{array}$$

where, for $a_{p,q}\in H^{2q-p}(X,q)$, *we use the shorthand notation* $\cup_k$*:*

$$\cup_k\left(\sum_{j+n=k} a_{i,j}\otimes a_{m,n}\right) = \sum_{j+n=k} -\frac{(k-1)!}{(j-1)!(n-1)!}\, a_{i,j}\cup a_{m,n}\,.$$

Definition 4.6.2 *The Chern character* $\mathrm{ch}_i : K_i(X)\to \bigoplus_{j\geq 0} H^{2j-i}(X,j)$ *is given by*

$$\mathrm{ch}_i(x) = \begin{cases} \sum_{j\geq 1} \frac{(-1)^{j-1}}{(j-1)!}\, c_{i,j}(x) & ,\ \text{if } i\geq 1 \\ \mathrm{ch}_{0,0}(x) + \sum_{j\geq 0} \frac{(-1)^{j-1}}{(j-1)!}\, \tilde{c}_{0,j}(x) & ,\ \text{if } i=0\,, \end{cases}$$

where $\mathrm{ch}_{0,0} : K_0(X) \xrightarrow{\text{rank}} H^0(X,\mathbf{Z}) \longrightarrow H^0(X,\bullet)$, *and where the* $\tilde{c}_{0,j}(x)$ *are defined by the relation*

$$\sum_{j\geq 1} \tilde{c}_{0,j}(x)\, t^j = \log\big(1+\sum_{j\geq 1} c_{0,j}(x)\, t^j\big)\,.$$

Corollary 4.6.1 *(i) For* $x\in K_i(X)$ *and* $y\in K_j(X)$ *one has*

$$\mathrm{ch}_{i+j}(x\cdot y) = \mathrm{ch}_i(x)\cup \mathrm{ch}_j(y)\,.$$

(ii) Let $K_i^{(j)}(X) = \mathrm{Gr}^j_\gamma K_i(X)\otimes\mathbf{Q}$ *be the subspace of* $K_i(X)\otimes\mathbf{Q}$ *on which the Adams operations* ψ^k *act as multiplication by* k^j*, then*

$$\mathrm{ch}_i(K_i^{(j)}(X)) \subseteq H^{2j-i}(X,j)\,.$$

Remark 4.6.1 There exists also a relative version of the Chern character

$$\mathrm{ch}_i^Y : K_i^Y(X) \longrightarrow \bigoplus_{j\geq 0} H_Y^{2j-i}(X,j)\,,$$

for a closed subscheme Y of X in $\mathcal{V}_S$.

Remark 4.6.2 The Chern characters ch_i and ch_i^Y are ring homomorphisms.

4.7 Gillet's Riemann-Roch Theorem

We are now able to formulate H. Gillet's Riemann-Roch Theorem for higher algebraic K-theory (cf. [Gil]).

Theorem 4.7.1 (Gillet) *Let $(H^\bullet, H_\bullet)$ be a Poincaré duality theory on the category $\mathcal{V}_S$, and let $\mathcal{V}_{S,*}$ denote the category with the same objects as $\mathcal{V}_S$ but with morphisms, the projective morphisms in $\mathcal{V}_S$. Then there is a natural transformation $\tau_* = \oplus_{i\geq 0}\tau_i$ of covariant functors $\mathcal{V}_S \longrightarrow \mathcal{G}r\mathcal{A}b$,*

$$\tau_i : K_i'(X) \longrightarrow \bigoplus_{j\geq 0} H_{i-2j}(X,j)\otimes \mathbf{Q}\,,$$

which satisfies the following conditions:

(i) The following diagram commutes

$$\begin{array}{ccc} K_i(X)\otimes K_j'(X) & \xrightarrow{\quad\cap\quad} & K_{i+j}'(X) \\ \Big\downarrow \mathrm{ch}_i \,\Big|\, \tau_j & & \Big\downarrow \\ \bigoplus_{n-m=k} H^{2m-i}(X,m)\otimes H_{2n+j}(X,n) & \xrightarrow{\ \cap\ } & \bigoplus_{k\geq 0} H_{2k+i+j}(X,k)\,. \end{array}$$

(ii) For an open immersion $j : U \hookrightarrow X$ in $\mathcal{V}_S$ and $x \in K_i'(X)$ the following holds:

$$\tau_i(j^*(x)) = j^!\tau_i(x)\,.$$

(iii) If $X \in \mathcal{O}b(\mathcal{V}_S)$ is smooth, then the structure sheaf $\mathcal{O}_X$ defines an element $[\mathcal{O}_X] \in K_0'(X) \cong K_0(X)$, and

$$\tau_0([\mathcal{O}_X]) = \mathrm{td}(X)\cap \eta_X\,.$$

Thus τ_0 coincides with τ of the Riemann-Roch Theorem of Baum, Fulton and MacPherson.

(iv) If $X, Y \in \mathcal{O}(\mathcal{V}_S)$ and $x \in K_i'(X)$, $y \in K_j'(Y)$, then

$$\tau_{i+j}(x\times y) = \tau_i(x)\times \tau_j(y)\,,$$

where $\times$ denotes the external product in both K'-theory and homology.

Remark 4.7.1 Most of the time we write just ch for ch_i,... etc.

The following definition is due to A. Beilinson. It will play a major role in the sequel.

Definition 4.7.1 (Beilinson) *The* **absolute** *or* **motivic cohomology** *of X is defined by*

$$H^i_{\mathcal{M}}(X, \mathbf{Q}(j)) = K^{(j)}_{2j-i}(X) = \mathrm{Gr}^j_\gamma K_{2j-i}(X) \otimes \mathbf{Q}\,.$$

Remark 4.7.2 Bloch (cf. [Bl3]) proves an interesting relationship between his generalized Chow groups $CH^r(X,n)$ and Beilinson's motivic cohomology:

$$CH^j(X, 2j-i) \otimes \mathbf{Q} \xrightarrow{\sim} H^i_{\mathcal{M}}(X, \mathbf{Q}(j))\,.$$

According to Soulé's theorem 4.5.1 the $H^i_{\mathcal{M}}(X, \mathbf{Q}(j))$ are the cohomological part of a Poincaré duality theory, so there exists a Chern character

$$\mathrm{ch}_{\mathcal{M}} : K_i(X) \longrightarrow \bigoplus_{j \geq 0} H^{2j-i}_{\mathcal{M}}(X, \mathbf{Q}(j))\,,$$

and actually, tensoring with $\mathbf{Q}$,

$$\mathrm{ch}_{\mathcal{M}} \otimes \mathbf{Q} : K_i(X) \otimes \mathbf{Q} \longrightarrow \bigoplus_{j \geq 0} H^{2j-i}_{\mathcal{M}}(X, \mathbf{Q}(j))$$

gives the decomposition of $K_i(X) \otimes \mathbf{Q}$ into eigensubspaces of eigenvalues k^j under the Adams operations

$$\psi^k : K_i(X) \longrightarrow K_i(X)\,.$$

4.8 Motivic Cohomology

The $H^i_{\mathcal{M}}(X, \mathbf{Q}(j))$ are believed to have a universal nature in the following sense (cf. [BMS]).

Conjecture 4.8.1 (Beilinson, MacPherson, Schechtman) *There are complexes of sheaves* $\mathbf{Z}(j)^{\bullet}_{\mathcal{M}}$ *on* $X \in Ob(\mathcal{V}_S)$ *such that*

$$H^i_{\mathcal{M}}(X, \mathbf{Q}(j)) = \mathbf{H}^i(X, \mathbf{Z}(j)^{\bullet}_{\mathcal{M}}) \otimes \mathbf{Q},$$

and the direct sum $\bigoplus_{j\geq 0} \mathbf{H}^i(X, \mathbf{Z}(j)^{\bullet}_{\mathcal{M}})$ *is universal with respect to refinements of the usual cohomology groups* $H^i(X)$, *i.e. it maps to other ones.*
Also, there should be 'cycle maps'

$$\mathbf{H}^i(X, \mathbf{Z}(j)^{\bullet}_{\mathcal{M}}) \longrightarrow H^i(X),$$

for any cohomology theory $H^i(X)$.

Example 4.8.1 For varieties X over $\mathbf{C}$, $H^i(X)$ may be singular cohomology and Deligne-Beilinson cohomology $H^i_{\mathcal{D}}(X, \mathbf{Z}(j))$ is such a refinement. The 'cycle map' should factor as follows:

$$\mathbf{H}^i(X, \mathbf{Z}(j)^{\bullet}_{\mathcal{M}}) \longrightarrow H^i_{\mathcal{D}}(X, \mathbf{Z}(j)) \longrightarrow H^i(X).$$

The first three $\mathbf{Z}(j)^{\bullet}_{\mathcal{M}}$ should be as follows:
(*i*) $\mathbf{Z}(0)^{\bullet}_{\mathcal{M}} \simeq \mathbf{Z}$, the constant sheaf, placed in degree zero.
(*ii*) $\mathbf{Z}(1)^{\bullet}_{\mathcal{M}} \simeq \mathbf{G}_m[-1] \simeq \mathcal{O}^*$, placed in degree one, so that $\mathbf{Z}(1)^{\bullet}_{\mathcal{M}}$ is quasi-isomorphic with Deligne's cohomology sheaf $\mathbf{Z}(1)_{\mathcal{D}}$.
(*iii*) $\mathbf{Z}(2)^{\bullet}_{\mathcal{M}} \simeq (\mathbf{Z}(\mathcal{O}^{**})/R \xrightarrow{d} \wedge^2\mathcal{O}^*)$, the Bloch-Suslin complex, placed in degrees one and two. Here $\wedge^2\mathcal{O}^*$ is the exterior power of $\mathcal{O}^*$ as a $\mathbf{Z}$-module with multiplication instead of addition. $\mathcal{O}^{**}$ denotes the set of elements $t \in \mathcal{O}^*$ such that also $1 - t \in \mathcal{O}^*$, and d takes t into $t \wedge (1-t)$. R is the subgroup generated by the expressions

$$[x] + [y] + [\frac{1-x}{1-xy}] + [(1-xy)] + [\frac{1-y}{1-xy}].$$

These occur in the functional equation of the dilogarithm, cf. next chapter.

For the category $\mathcal{V} = \mathcal{S}ch/\mathbf{C}$ (or $\mathcal{S}ch/\mathbf{R}$) of schemes (of finite type) over $\mathbf{C}$ (or $\mathbf{R}$) Beilinson's ideas take the following form.

First, the natural setting is in terms of **derived categories**, i.e. categories whose objects are complexes (of sheaves) and whose morphisms can be defined along a now standard procedure, first developed by J-L. Verdier (1935–1989) (cf. [Ve]). According to Deligne's 'dictionnaire heuristique' the role of sheaves on Spec($\mathbf{C}$) in an arithmetic context is played by **(mixed) Hodge structures**. The category $\mathcal{H}$ of (mixed) Hodge structures on $\mathcal{V}$ is abelian, hence by general principles (cf. [Ve]) admits a derived category $\mathrm{D}^b(\mathcal{H})$ whose elements are bounded complexes of (mixed) Hodge structures. For any $X \in \mathcal{O}b(\mathcal{V})$ Beilinson constructs a canonical object $\underline{\mathrm{R}\Gamma}(X, \mathbf{Z}) \in \mathcal{O}b(\mathrm{D}^b(\mathcal{H}))$, such that, for all $j \in \mathbf{Z}$,

$$\mathrm{R}\Gamma_{\mathcal{H}}(X, \mathbf{Z}(j)) = \mathrm{RHom}_{\mathrm{D}^b(\mathcal{H})}(\mathbf{Z}(0), \underline{\mathrm{R}\Gamma}(X, \mathbf{Z})(j)),$$

where (j) means Tate twist in the derived category $\mathrm{D}^b(\mathcal{H})$, can be considered as a natural cochain complex on X, the so-called **absolute Hodge complex** on X. Thus, writing

$$\underline{\mathrm{H}}^\bullet : \mathrm{D}^b(\mathcal{H}) \longrightarrow \mathcal{H}$$

for the cohomology functor, one finds that $\underline{\mathrm{H}}^\bullet\underline{\mathrm{R}\Gamma}(X, \mathbf{Z})(\star)$ is just the singular cohomology $H^\bullet(X, \mathbf{Z}(\star))$ with its mixed Hodge structure.

Definition 4.8.1 (Beilinson) *The* **absolute Hodge cohomology** *of X is*

$$H^\bullet_{\mathcal{H}}(X, \mathbf{Z}(\star)) = H^\bullet(\mathrm{R}\Gamma_{\mathcal{H}}(X, \mathbf{Z}(\star))) =$$
$$= H^\bullet\mathrm{RHom}_{\mathrm{D}^b(\mathcal{H})}(\mathbf{Z}(0), \underline{\mathrm{R}\Gamma}(X, \mathbf{Z})(\star)).$$

For $i \leq 2j$, $H^i_{\mathcal{H}}(X, \mathbf{Z}(j))$ coincides with $H^i_{\mathcal{D}}(X, \mathbf{Z}(j))$. Absolute Hodge cohomology admits a homological counterpart to give a Poincaré duality theory. It generalizes Deligne-Beilinson cohomology in the sense that it takes the weight filtration into account. More precisely, one has an exact sequence

$$0 \longrightarrow W_{2j}H^i(X, \mathbf{C})/[W_{2j}H^i(X, \mathbf{Q}(j)) \oplus (F^j \cap W_{2j})H^i_{DR}(X)] \longrightarrow$$
$$\longrightarrow H^{i+1}_{\mathcal{H}}(X, \mathbf{Z}(j)) \longrightarrow W_{2j}H^i(X, \mathbf{Q}(j)) \cap F^j H^i_{DR}(X) \longrightarrow 0 \tag{4.4}$$

Beilinson calls the formerly defined Deligne-Beilinson cohomology **weak absolute Hodge cohomology**. We come back to absolute Hodge cohomology and its relation to a generalized Hodge Conjecture in Chapter 8. A detailed account can be found in [Be3].

Secondly, Beilinson conjectures the universality of the motivic cohomology $H^{\bullet}_{\mathcal{M}}(X, \mathbf{Q}(\star))$ in the following way, already alluded to in Conjecture 4.8.1

$$H^{\bullet}_{\mathcal{M}}(X, \mathbf{Q}(\star)) = \mathbf{H}^{\bullet}(X, \mathbf{Z}(\star)^{\bullet}_{\mathcal{M}}) \otimes \mathbf{Q},$$

where $\mathbf{Z}(\star)^{\bullet}_{\mathcal{M}}$ is an element (i.e. a complex) of the derived category of a conjectural category of so-called **mixed motives** $\mathcal{MM}$, in other words

$$\mathbf{H}^{\bullet}(X, \mathbf{Z}(\star)^{\bullet}_{\mathcal{M}}) = \mathrm{Hom}_{\mathrm{D}(\mathcal{MM})}(\mathbf{Z}, \mathbf{Z}(\star))[\bullet] = \mathrm{Ext}^{\bullet}_{\mathcal{MM}}(\mathbf{Z}, \mathbf{Z}(\star)).$$

Thus $\mathrm{D}(\mathcal{MM}) = \mathrm{D}(\mathcal{MM})(X)$ is a certain triangulated category of 'motivic sheaves' on X. This triangulated category should satisfy the formalism of a mixed sheaf theory, i.e. it should admit the usual operations $\mathrm{R}f_*$, $\mathrm{R}f_!$, f^*, $\mathrm{R}f^!$, $\otimes$, $\mathrm{R}\underline{\mathrm{Hom}}$, Verdier duality, and, in a suitable sense, vanishing cycles. The formalism of **mixed perverse sheaves** should apply to $\mathrm{D}(\mathcal{MM})(X)$, (cf. [BBD]).

It is also conjectured that there is a natural realization functor

$$r_{\mathcal{H}} : \mathrm{D}(\mathcal{MM})(X) \longrightarrow \mathrm{D}(\mathcal{H})(X)$$

from motivic sheaves to mixed Hodge structures on X, inducing so-called **regulator maps**

$$r : H^{\bullet}_{\mathcal{M}}(X, \mathbf{Q}(\star)) \longrightarrow H^{\bullet}_{\mathcal{H}}(X, \mathbf{Q}(\star)).$$

These regulator maps should be determined by the corresponding Chern characters, to make the following diagram commute:

$$\begin{array}{ccc} K_i(X) & \xrightarrow{\mathrm{ch}_{\mathcal{M}}} & H^{\bullet}_{\mathcal{M}}(X, \mathbf{Q}(\star)) \\ \| & & \downarrow r \\ K_i(X) & \xrightarrow{\mathrm{ch}_{\mathcal{H}}} & H^{\bullet}_{\mathcal{H}}(X, \mathbf{Q}(\star)). \end{array}$$

Chapter 5

Regulators, Deligne's conjecture and Beilinson's first conjecture

In this chapter we describe two examples of maps from algebraic K-theory to Deligne-Beilinson cohomology that can be considered as a first motivation for Beilinson's conjectures. These conjectures are then formulated in such a way that they generalize, at the same time, a conjecture of Deligne on the values of L-functions of motives at so-called critical points. We will state the conjectures only for smooth projective varieties defined over the rational numbers, but it should be observed that almost everything can be formulated for motives over arbitrary number fields, the statements becoming just more complicated.

5.1 Borel's Regulator

The first example was inspired by work of A. Borel on arithmetic groups (cf. [Bo1] and [Bo2]). For a number field K Borel calculated the rank of the K-groups of the ring of integers $\mathcal{O}_K$.

Theorem 5.1.1 *(i)* $\dim_{\mathbf{Q}} K_{2i}(\mathcal{O}_K) \otimes \mathbf{Q} = 0$, $i = 1, 2, \ldots$

(ii) $\dim_{\mathbf{Q}} K_{2i-1}(\mathcal{O}_K) \otimes \mathbf{Q} = \begin{cases} r_1 + r_2, & i \geq 2 \textit{ and odd} \\ r_2, & i \geq 2 \textit{ and even}. \end{cases}$

Here, as always, r_1 (resp. r_2) denotes the number of real (resp. complex conjugate) embeddings of K.

Remark 5.1.1 For $K_1(\mathcal{O}_K)$ one has Dirichlet's Unit Theorem 1.3.1, saying in this language that $\dim_{\mathbf{Q}} K_1(\mathcal{O}_K) \otimes \mathbf{Q} = r_1 + r_2 - 1$.

Corollary 5.1.1 *For all integers $m = 1, 2, \ldots$, one has:*

$$\mathrm{ord}_{s=1-m}\, \zeta_K(s) = \dim_{\mathbf{Q}} K_{2m-1}(\mathcal{O}_K) \otimes \mathbf{Q}\,.$$

Also, using group cohomology, Borel was led to define so-called **regulator maps**

$$r_{i,Borel} : K_{2i-1}(K) \otimes \mathbf{Q} \longrightarrow H^1_{\mathcal{D}}(X_{/\mathbf{R}}, \mathbf{R}(i)),$$

where $X = \mathrm{Spec}(K)$. Now, because $K_j(K) \otimes \mathbf{Q} \simeq K_j(\mathcal{O}_K) \otimes \mathbf{Q}$, $j > 1$, this means that one has regulator maps, also written $r_{i,Borel}$,

$$r_{i,Borel} : K_{2i-1}(\mathcal{O}_K) \otimes \mathbf{Q} \longrightarrow H^1_{\mathcal{D}}(X_{/\mathbf{R}}, \mathbf{R}(i)).$$

These regulator maps are injective homomorphisms and they define a $\mathbf{Q}$–structure on Deligne cohomology. The general Deligne cohomology $H^i_{\mathcal{D}}(X_{/\mathbf{R}}, \mathbf{R}(p))$ has a natural $\mathbf{Q}$–structure as follows: From the basic exact sequence (3.3)

$$0 \to F^p H^{i-1}_{DR}(X_{/\mathbf{R}}) \longrightarrow H^{i-1}(X_{/\mathbf{R}}, \mathbf{R}(p-1)) \longrightarrow H^i_{\mathcal{D}}(X_{/\mathbf{R}}, \mathbf{R}(p)) \to 0,$$

one gets for the maximal exterior powers, denoted by det,

$$\det{}_{\mathbf{R}} H^i_{\mathcal{D}}(X_{/\mathbf{R}}, \mathbf{R}(p)) =$$

$$= \det{}_{\mathbf{R}} H^{i-1}(X_{/\mathbf{R}}, \mathbf{R}(p-1)) \otimes (\det{}_{\mathbf{R}} F^p H^{i-1}_{DR}(X_{/\mathbf{R}}))^{\vee} \quad (5.1)$$

where $^{\vee}$ denotes the dual. But the right hand side admits a $\mathbf{Q}$–structure

$$\Lambda_{p,i} = \det{}_{\mathbf{Q}} H^{i-1}(X_{/\mathbf{R}}, \mathbf{Q}(p-1)) \otimes (\det{}_{\mathbf{Q}} F^p H^{i-1}_{DR}(X_{/\mathbf{Q}}))^{\vee}.$$

Here $H^i_{DR}(X_{/\mathbf{Q}})$ means the algebraic de Rham cohomology. The $\Lambda_{i,p}$ is by definition the natural $\mathbf{Q}$–structure on Deligne cohomology. For any $\mathbf{Q}$–structure $\Phi_{p,i}$ of $H^i_{\mathcal{D}}(X_{/\mathbf{R}}, \mathbf{R}(p))$ its **volume** $\mathrm{Vol}(\Phi_{p,i})$ is given by the

Definition 5.1.1 $\det_{\mathbf{Q}} \Phi_{p,i} = \mathrm{Vol}(\Phi_{p,i}) . \Lambda_{p,i}$.

Coming back to Borel's regulator maps, one defines **Borel's regulators** $R_{i,Borel}$ as the volumes of the $\mathbf{Q}$-structures defined by the $r_{i,Borel}$. Borel's fundamental result is the following theorem.

Theorem 5.1.2 (Borel) $R_{m,Borel} \approx_{\mathbf{Q}^*} \zeta_K^*(1-m)$, $m \geq 2$.

Here, as in the sequel, $\zeta_K^*(t)$, $t \in \mathbf{Z}$, denotes the first non-zero coefficient in the Taylor series expansion of $\zeta_K(s)$ at the point $s = t$.

Remark 5.1.2 For $m = 1$ Dedekind's Class Number Formula says that $R \approx_{\mathbf{Q}^*} \zeta_K^*(0)$, where R is the classical regulator.

5.2 Beilinson's Regulator

Beilinson's approach to regulator maps is apparently different. Let K be a (finite) number field and write $X = \mathrm{Spec}(K)$. The explicit construction of Beilinson's **regulator maps**, written $r_{i,\mathcal{D}}$, or simply $r_\mathcal{D}$ when no confusion is likely to arise,

$$r_{i,\mathcal{D}} : H^1_{\mathcal{M}}(X, \mathbf{Q}(i)) \longrightarrow H^1_{\mathcal{D}}(X, \mathbf{R}(i)) = \left[\bigoplus_{\sigma: K \longrightarrow \mathbf{C}} \mathbf{R}(i-1) \right]^{DR} =$$

$$= \left[\bigoplus_{\sigma: K \longrightarrow \mathbf{C}} \mathbf{R}(i-1) \right]^{\mathrm{Gal}(\mathbf{C}/\mathbf{R})} = \begin{cases} \mathbf{R}^{r_1+r_2}, & \text{if } i \geq 1 \text{ and odd} \\ \mathbf{R}^{r_2}, & \text{if } i \geq 2 \text{ and even}, \end{cases}$$

is based on the evaluation map

$$e : \mathrm{Spec}(\mathbf{C}) \times \mathrm{B_{\bullet}GL_N}(\mathbf{C}) \longrightarrow \mathrm{B_{\bullet}GL_{N/\mathbf{C}}},$$

where $\mathrm{B_{\bullet}GL_N}(\mathbf{C})$ is considered as the disjoint union $\coprod_x \mathrm{Spec}(\mathbf{C})$ over sections $x : \mathrm{Spec}(\mathbf{C}) \longrightarrow \mathrm{B_{\bullet}GL_{N/\mathbf{C}}}$ of $\mathrm{B_{\bullet}GL_{N/\mathbf{C}}} \longrightarrow \mathrm{Spec}(\mathbf{C})$, the structure map. For the ith universal Chern class

$$c_i^{(N)} \in H^{2i}_{\mathcal{D}}(\mathrm{B_{\bullet}GL_{N/\mathbf{C}}}, \mathbf{Q}(i)) \xrightarrow{\sim} H^{2i}(\mathrm{B_{\bullet}GL_{N/\mathbf{C}}}, \mathbf{Q}(i)),$$

(cf. the exact sequences (a) and (b) in Proposition 3.2.2 with $\mathbf{R}$ replaced by $\mathbf{Q}$) one obtains, using the Künneth formula,

$$e^*(c_i^{(N)}) \in H^{2i}_{\mathcal{D}}(\mathrm{Spec}(\mathbf{C}) \times \mathrm{B_{\bullet}GL_N}(\mathbf{C}), \mathbf{Q}(i)) \longrightarrow$$

$$\longrightarrow H_{\mathcal{D}}^{2i}(\mathrm{Spec}(\mathbf{C}) \times \mathrm{B}_{\bullet}\mathrm{GL}_N(\mathbf{C}), \mathbf{R}(i)) \xrightarrow{\sim}$$
$$\xrightarrow{\sim} H_{\mathcal{D}}^{1}(\mathrm{Spec}(\mathbf{C}), \mathbf{R}(i)) \otimes H^{2i-1}(\mathrm{B}_{\bullet}\mathrm{GL}_N(\mathbf{C}), \mathbf{R}) \xrightarrow{\sim}$$
$$\xrightarrow{\sim} \mathbf{R}(i-1) \otimes H^{2i-1}(\mathrm{B}_{\bullet}\mathrm{GL}_N(\mathbf{C}), \mathbf{R}) \xrightarrow{\sim} H^{2i-1}(\mathrm{B}_{\bullet}\mathrm{GL}_N(\mathbf{C}), \mathbf{R}(i-1)) =$$
$$= H^{2i-1}(\mathrm{GL}_N(\mathbf{C}), \mathbf{R}(i-1)) = \mathrm{Hom}(H_{2i-1}(\mathrm{GL}_N(\mathbf{C}), \mathbf{Z}), \mathbf{R}(i-1)) =$$
$$= \mathrm{Hom}(H_{2i-1}(\mathrm{GL}(\mathbf{C}), \mathbf{Z}), \mathbf{R}(i-1)) \qquad (5.2)$$

for N large enough. The element of $\mathrm{Hom}(H_{2i-1}(\mathrm{GL}(\mathbf{C}), \mathbf{Z}), \mathbf{R}(i-1))$ is DR-invariant. Using the Hurewicz map, one gets a map

$$\phi_i : K_{2i-1}(\mathbf{C}) = \pi_{2i-1}(\mathrm{BGL}(\mathbf{C})^+) \longrightarrow H_{2i-1}(\mathrm{GL}(\mathbf{C}), \mathbf{Z}) \longrightarrow$$
$$\longrightarrow \mathbf{R}(i-1) \simeq H_{\mathcal{D}}^{1}(\mathrm{Spec}(\mathbf{C}), \mathbf{R}(i)).$$

For the number field K one has

$$X_{/\mathbf{C}} = \mathrm{Spec}(K) \otimes \mathbf{C} = \prod_{\sigma : K \longrightarrow \mathbf{C}} \mathrm{Spec}(\mathbf{C})$$

and the **regulator map** $r_{i,\mathcal{D}}$ is defined as

$$\frac{(-1)^{i-1}}{(i-1)!}\left[\prod_{\sigma : K \longrightarrow \mathbf{C}} \phi_i \circ \sigma^*\right]^{DR} : K_{2i-1}(X) \longrightarrow \left[\prod_{\sigma} K_{2i-1}(\mathbf{C})\right]^{DR} \longrightarrow$$
$$\longrightarrow H_{\mathcal{D}}^{1}(X_{/\mathbf{C}}, \mathbf{R}(i))^{DR} = \left[\bigoplus_{\sigma} H_{\mathcal{D}}^{1}(\mathrm{Spec}(\mathbf{C}), \mathbf{R}(i))\right]^{DR}. \qquad (5.3)$$

Composing with $K_{2i-1}(X_{\mathbf{Z}}) = K_{2i-1}(\mathcal{O}_K) \longrightarrow K_{2i-1}(X)$, one obtains a map

$$r_{i,\mathcal{D}} : K_{2i-1}(X_{\mathbf{Z}}) \longrightarrow H_{\mathcal{D}}^{1}(X_{/\mathbf{R}}, \mathbf{R}(i)).$$

Tensoring with $\mathbf{Q}$ and using the compatibility of the Adams operations and Chern classes (or Chern character) one finally obtains **Beilinson's regulator maps** $r_{i,\mathcal{D}}$ or simply $r_{\mathcal{D}}$,

$$r_{\mathcal{D}} : H_{\mathcal{M}}^{1}(X, \mathbf{Q}(i)) \longrightarrow H_{\mathcal{D}}^{1}(X_{/\mathbf{R}}, \mathbf{R}(i)),$$

or in a form more appropriate for the future formulation of Beilinson's conjectures, taking care of results of Bloch and Grayson to be discussed below, one has Beilinson's regulator map

$$r_{\mathcal{D}} : H_{\mathcal{M}}^{1}(X_{\mathbf{Z}}, \mathbf{Q}(i)) \longrightarrow H_{\mathcal{D}}^{1}(X_{/\mathbf{R}}, \mathbf{R}(i)). \qquad (5.4)$$

Remark 5.2.1 Actually, $K_{2i-1}(X_{\mathbf{Z}}) \otimes \mathbf{Q} \simeq H^1_{\mathcal{M}}(X_{\mathbf{Z}}, \mathbf{Q}(i))$ and $r_{\mathcal{D}}$ is just the Chern character.

An important result, due to Beilinson (cf. [Bel] and for a detailed account [Rap]), is the following theorem.

Theorem 5.2.1 *Up to a non-zero rational number Borel's regulator maps and Beilinson's regulator maps coincide, in other words*

$$r_{i,Borel} \approx_{\mathbf{Q}^*} r_{i,\mathcal{D}}, \quad i \geq 2.$$

Borel's results may now be translated in terms of motivic (or absolute) cohomology and regulator maps as follows:

Theorem 5.2.2 (Borel-Beilinson) *Let K be a (finite) number field with ring of integers $\mathcal{O}_K$, and let $X = \mathrm{Spec}(K)$ and $X_{\mathbf{Z}} = \mathrm{Spec}(\mathcal{O}_K)$. Then, for an integer $m \geq 2$,*

(i) $\mathrm{ord}_{s=1-m}\, \zeta_K(s) = \dim_{\mathbf{Q}} K_{2m-1}(\mathcal{O}_K) \otimes \mathbf{Q} = \dim_{\mathbf{Q}} H^1_{\mathcal{M}}(X_{\mathbf{Z}}, \mathbf{Q}(m))$.

(ii) $r_{\mathcal{D}} : H^1_{\mathcal{M}}(X_{\mathbf{Z}}, \mathbf{Q}(m)) \longrightarrow H^1_{\mathcal{D}}(X_{/\mathbf{R}}, \mathbf{R}(m))$ *defines a $\mathbf{Q}$-structure on $H^1_{\mathcal{D}}(X_{/\mathbf{R}}, \mathbf{R}(m))$ with volume $R_{\mathcal{D}}$, such that*

$$R_{\mathcal{D}} \approx_{\mathbf{Q}^*} \zeta_K^*(1-m).$$

$R_{\mathcal{D}}$ is called the **regulator** and is sometimes denoted by $\det(r_{\mathcal{D}})$.

Remark 5.2.2 For $m = 1$, (i) remains correct, and 'thickening' $r_{\mathcal{D}}$ to

$$r'_{\mathcal{D}} : \mathbf{Q} \oplus H^1_{\mathcal{M}}(X_{\mathbf{Z}}, \mathbf{Q}(1)) = (\mathbf{Z} \oplus \mathcal{O}_K^*) \otimes \mathbf{Q} \longrightarrow H^1_{\mathcal{D}}(X_{/\mathbf{R}}, \mathbf{R}(1)) = \mathbf{R}^{r_1+r_2}$$

by embedding $\mathbf{Z}$ diagonally (cf. Chapter 1) and defining $R_{\mathcal{D}} = R$, the classical regulator, (ii) is also true. In the sequel we will assume this thickening of $r_{\mathcal{D}}$ and this definition of $R_{\mathcal{D}}$, so that the statements are valid for all $m \geq 1$.

5.3 Special Cases and Zagier's Conjecture

The cases $m = 2$ and $m = 3$ deserve special attention. We know that the Bloch-Suslin complex $\mathbf{Z}(2)^{\bullet}_{\mathcal{M}}$ is a candidate for motivic cohomology

$$H^1_{\mathcal{M}}(X, \mathbf{Q}(2)) = \mathbf{H}^1(X, \mathbf{Z}(2)^{\bullet}_{\mathcal{M}}) \otimes \mathbf{Q}.$$

We discuss briefly how this may lead to an explicit description of Borel's, and by Beilinson's result, also of Beilinson's regulator map, up to a non-zero rational multiple.

To this end, let $Li_0(z) = \frac{z}{(1-z)}$, $z \in \mathbf{C}$, and define, as usual, the higher polylogarithms $Li_k(z)$ inductively by

$$Li_k(z) = \int_0^z Li_{k-1}(t)\frac{dt}{t}\,,$$

where the path of integration avoids the cut from 1 to $+\infty$. $Li_k(z)$ is the analytic continuation of the series

$$Li_k(z) = \sum_{n=1}^{\infty} \frac{z^n}{n^k}\,, \qquad |z| \leq 1\,.$$

Thus $Li_1(z) = -\log(1-z)$ and $Li_2(z)$ is the dilogarithm function of z. Next, one defines the function $D_2 : \mathbf{P}^1_{\mathbf{C}} \longrightarrow \mathbf{C}$ by

$$D_2(z) = \begin{cases} \Im(Li_2(z)) + \arg(1-z)\cdot\log|z| & , \text{ if } z \in \mathbf{P}^1_{\mathbf{C}} \setminus \{0,1,\infty\} \\ 0 & , \text{ if } z = 0,\ 1 \text{ or } \infty\,. \end{cases}$$

This function enjoys several remarkable properties, e.g.

(*i*) D_2 is continuous on all of $\mathbf{P}^1_{\mathbf{C}}$, and real analytic on $\mathbf{P}^1_{\mathbf{C}} \setminus \{0,1,\infty\}$.

(*ii*) $D_2(z^{-1}) = -D_2(z) = D_2(1-z) = D_2(\frac{z}{z-1})$, for any $z \in \mathbf{P}^1_{\mathbf{C}}$.

(*iii*) $D_2(\bar{z}) = -D_2(z)$, in particular $D_2(x) = 0$, $x \in \mathbf{R}$.

(*iv*) For all $x, y \in \mathbf{P}^1_{\mathbf{C}}$ one has

$$D_2(x) + D_2(y) + D_2(\frac{1-x}{1-xy}) + D_2(1-xy) + D_2(\frac{1-y}{1-xy}) = 0\,.$$

Property (*iv*) reflects Abel's (and Spence's) functional equation for the dilogarithm, and the expression of the left hand side of (*iv*) is used in the definition of the Bloch-Suslin complex $\mathbf{Z}(2)^{\bullet}_{\mathcal{M}}$, cf. Chapter 4. With an eye to Zagier's work on regulators for number fields K, we recall the definition of the Bloch-Suslin complex for the zero dimensional case $X = \mathrm{Spec}(K)$, K a number field, cf. [Gon]. To this end, let $\mathbf{P}^1_K$ denote the projective line over K and define $Z(K) = \mathbf{Z}[\mathbf{P}^1_K \setminus \{0,1,\infty\}]$ as the free abelian group generated by symbols $[x]$, where $x \in \mathbf{P}^1_K \setminus \{0,1,\infty\}$. Then one may consider D_2 as a homomorphism

$$D_2 : Z(K) \longrightarrow \mathbf{R} \quad \text{by} \quad D_2\left(\sum_i n_i[x_i]\right) = \sum_i n_i D_2(x_i)\,,$$

where we consider the x_i as real or complex numbers after some embedding of K into $\mathbf{R}$ or $\mathbf{C}$. In the sequel we will sometimes tacitly assume such an embedding without mentioning it if its explicit form is irrelevant, and consider the polylogarithm functions as functions with a complex argument. Abel's functional equation implies that D_2 vanishes on the subgroup R_2 of Z generated by elements of the form

$$[x] + [y] + [\frac{1-x}{1-xy}] + [1-xy] + [\frac{1-y}{1-xy}]\,.$$

This fact can also be interpreted, in a more geometric way, as follows. For any 5-tuple of different points $x_0, \ldots, x_4$ in $\mathbf{P}^1_K$ we will write $R_2(x_0, \ldots, x_4)$ for the alternating sum of the cross-ratios $\{\cdot,\cdot,\cdot,\cdot\}$ of four of these points as follows

$$R_2(x_0, \ldots, x_4) = \sum_{i=0}^{4} (-1)^i [\{x_0, \ldots, \hat{x}_i, \ldots, x_4\}] \in Z(K)\,.$$

In fact, because of the PGL_2-invariance of the cross-ratio, this expression depends only on two variables. Then $R_2 \subset Z(K)$ is the subgroup generated by the elements $R_2(x_0, \ldots, x_4)$ for all choices of the $x_0, \ldots, x_4$. In this context, Abel's functional equation leads to the nice formula

$$\sum_{i=0}^{4} (-1)^i D_2(\{z_0, \ldots, \hat{z}_i, \ldots, z_4\}) = 0 \qquad (5.5)$$

which can be interpreted in terms of the geometry of hyperbolic 3-space $\mathcal{H}_3$ represented as $\mathbf{C} \times \mathbf{R}^+$ with the standard hyperbolic metric such that the geodesics are either vertical lines or half-circles in vertical planes with endpoints in $\mathbf{C} \times \{0\}$. An **ideal hyperbolic 3-simplex** Δ is a tetrahedron whose vertices are all in $\partial\mathcal{H}_3 = \mathbf{C} \cup \infty = \mathbf{P}^1(\mathbf{C})$. It is well-known that the volume of such a Δ, with vertices z_0, z_1, z_2, z_3, is given by

$$\mathrm{Vol}(\Delta) = D_2(\{z_0, z_1, z_2, z_3\}),$$

and, in particular, when the vertices are ∞, 0, 1 and z, respectively, one has $\mathrm{Vol}(\Delta) = D_2(z)$. Now equation (5.5) just says that the five tetrahedra spanned by four of the points z_i, $i = 0, \dots, 4$, at the time add up algebraically to the zero 3-cycle.
Returning to what we were after, define

$$B_2(K) = Z(K)/R_2.$$

The Bloch-Suslin complex $\mathbf{B}_2^\bullet(K) = \mathbf{Z}(2)^\bullet_{\mathcal{M}}$ is defined as the following complex, concentrated in degrees one and two:

$$\mathbf{B}_2^\bullet(K) : B_2(K) \xrightarrow{d} \wedge^2 K^*,$$

where $d\,[x] = x \wedge (1-x)$. The definition of d is also suggested by hyperbolic geometry. Namely, it can be shown that any complete oriented hyperbolic 3-manifold M with finite volume admits a finite triangulation by ideal hyperbolic 3-simplices, say $\Delta_1, \dots, \Delta_n$. After an isometry of $\mathcal{H}_3$, if necessary, we may assume that the vertices of Δ_i are at ∞, 0, 1 and z_i, $i = 1, \dots, n$. Thus

$$\mathrm{Vol}(M) = \sum_{i=1}^{n} \mathrm{Vol}(\Delta_i) = \sum_{i=1}^{n} D_2(z_i).$$

One verifies that this is well defined, i.e. does not depend on the order in which the vertices of the Δ_i were sent to ∞, 0, 1 and z_i. The z_i of the triangulation of a complete hyperbolic 3-manifold satisfy the remarkable relation

$$\sum_{i=1}^{n} z_i \wedge (1 - z_i) = 0 \qquad (5.6)$$

in the abelian group $\wedge^2\mathbf{C}^*$, i.e. the set of all formal linear combinations $x \wedge y$, $x, y \in \mathbf{C}\setminus\{0,1\}$, subject to the relations $x \wedge x = 0$ and $(x_1 x_2)\wedge y = x_1 \wedge y + x_2 \wedge y$. Equation (5.6) suggests the definition of the map d in the Bloch-Suslin complex.

Definition 5.3.1 *The* Bloch group $\mathcal{B}_2(K)$ *is defined as:*

$$\mathcal{B}_2(K) = \mathbf{H}^1(\mathrm{Spec}(K), \mathbf{Z}(2)^\bullet_{\mathcal{M}}) = \mathbf{H}^1(\mathrm{Spec}(K), \mathbf{B}^\bullet_2(K)) = \mathrm{Ker}(d).$$

It follows immediately from this definition that there is a map

$$\rho_{2,K} : \mathcal{B}_2(K) \longrightarrow \mathbf{R}^{r_2},$$

where r_2 is the number of (pairs of conjugate) complex embeddings σ of K, given by

$$\rho_{2,K}\left(\sum_i n_i[x_i]\right) = \left(\sum_i n_i D_2(\sigma(x_i))\right)_\sigma .$$

The real embeddings do not occur in this situation because $D_2(x) = 0$ for real x. Bloch and Suslin (cf. [Su]) proved that there is an isomorphism

$$\mathcal{B}_2(K) \otimes \mathbf{Q} \xrightarrow{\sim} K_3(K) \otimes \mathbf{Q} = K_3(\mathcal{O}_K) \otimes \mathbf{Q},$$

such that the composition with the Borel regulator map $r_{\mathcal{D}} = r_{2,\mathcal{D}} = r_{2,Borel}$ is just $\rho_{2,K} \otimes \mathbf{Q}$. This implies, in particular, that $\mathcal{B}_2(K)$ has rank r_2. The following result is due to D. Zagier [Za1]:

Theorem 5.3.1 *There are linearly independent elements $[x_i] \in \mathcal{B}_2(K)$, $i = 1, 2, \ldots, r_2$, such that*

$$\zeta_K(s)_{s=2} \approx_{\mathbf{Q}^*} \frac{\pi^{2(r_1+r_2)}}{\sqrt{|D(K)|}} \det\left(D_2(\sigma(x_i))\right)_{\sigma,i} .$$

The case $m = 3$ is equally interesting, and it is, by a recent result of A. Goncharov (cf. [Gon]), established. It involves the **trilogarithm** Li_3. To get 'clean' functional equations Zagier and Goncharov introduce the function[1]

$$D_3(z) = \Re\left\{Li_3(z) - \log|z| \cdot Li_2(z) + \frac{1}{3}\log^2|z| \cdot Li_1(z)\right\} .$$

The functional equations alluded to above are:

[1]Both authors use different notations for our D_3.

(*i*) D_3 is real analytic on $\mathbf{P}^1_{\mathbf{C}} \setminus \{0, \infty\}$ except at $z = 1$, and $D_3(1) = \zeta(3) = \zeta_{\mathbf{Q}}(3)$, Riemann's zeta-function at the point $s = 3$.

(*ii*) $D_3(z) = D_3(\frac{1}{z})$ for all $z \in \mathbf{P}^1_{\mathbf{C}} \setminus \{0, \infty\}$.

(*iii*) $D_3(z^2) = 4\,(D_3(z) + D_3(-z))$ for all $Z \in \mathbf{P}^1_{\mathbf{C}} \setminus \{0, \infty\}$.

(*iv*) $D_3(z) + D_3(1-z) + D_3(1-\frac{1}{z}) = D_3(1) = \zeta(s)_{s=3}$ for all $z \in \mathbf{P}^1_{\mathbf{C}} \setminus \{0, 1, \infty\}$.

(*v*) For $x, y \in \mathbf{P}^1_{\mathbf{C}} \setminus \{0, 1, \infty\}$ there is the 2-variable functional equation, due to Spence (1809) and Kummer (1840):

$$D_3(\frac{x(1-y)^2}{y(1-x)^2}) + D_3(xy) + D_3(\frac{x}{y}) - 2D_3(\frac{x(1-y)}{y(1-x)}) - 2D_3(\frac{x(1-y)}{x-1})$$

$$-2D_3(\frac{y(1-x)}{y-1}) - 2D_3(\frac{1-y}{1-x}) - 2D_3(x) - 2D_3(y) + 2D_3(1) = 0\,.$$

Goncharov [Gon] gives a geometric interpretation of this functional equation. It involves a special configuration of seven points x_1, x_2, x_3, y_1, y_2, y_3 and z in the projective plane $\mathbf{P}^2_K$ for the number field K. The x_i are the vertices of a triangle, the y_j are distributed over the three sides of this triangle, and z is in generic position. Taking all kinds of configurations of four points on a line obtained by projection of a 4-tuple of the x_i, y_j, z with center of projection a fifth point of the seven points, Goncharov takes a suitable combination $R_3(x_i, y_j, z)$ of the cross-ratios of these 4-tuples, considered as an element of $Z(K)$, and shows that

$$D_3(R_3(x_i, y_j, z)) = 0\,.$$

Here D_3 is considered as a homomorphism, just as D_2,

$$D_3 : \sum_i n_i[x_i] \mapsto \sum_i n_i D_3(x_i)\,, \quad x_i \in \mathbf{P}^1_K \setminus \{0, 1, \infty\}\,.$$

In case one of the x_i, one of the y_j and z are collinear, the 7-point configuration depends on only two parameters and the vanishing of $D_3(R_3(x_i, y_j, z))$ corresponds to the Spence-Kummer functional equation for D_3. It is conjectured that any functional equation for D_3 can be deduced from this geometric construction, and (*i*), (*ii*), (*iii*) and (*iv*) above. To apply the trilogarithm to regulator problems for the number field K, still following Goncharov, let $R_3(K)$ be the subgroup of $Z(K)$ generated by $R_3(x_i, y_j, z)$ and elements of the form

$$[x]-[\frac{1}{x}],\ [x^2]-4([x]+[-x]),\quad \text{and}$$

$$([x]+[1-x]+[1-\frac{1}{x}])-([y]+[1-y]+[1-\frac{1}{y}]),\ x,y\in \mathbf{P}^1_K\setminus\{0,1,\infty\}.$$

Now define the Goncharov complex $\mathbf{G}_3^\bullet(K) = \mathbf{Z}(3)^\bullet_{\mathcal{M}}$ as the following complex, placed in degrees one, two and three,

$$\mathbf{G}_3^\bullet(K) : Z(K)/R_3 \xrightarrow{d_1} B_2(K)\otimes K^* \xrightarrow{d_2} \wedge^3 K^*,$$

with $d_2\,[x]\otimes y = x\wedge(1-x)\wedge y$ and $d_1\,[x] = [x]\otimes x$. One is led to define another Bloch group:

Definition 5.3.2 *The* Bloch group $\mathcal{B}_3(K)$ *is defined as*

$$\mathcal{B}_3(K) = \mathbf{H}^1(\mathrm{Spec}(K), \mathbf{Z}(3)^\bullet_{\mathcal{M}}) = \mathbf{H}^1(\mathrm{Spec}(K), \mathbf{G}_3^\bullet) = \mathrm{Ker}(d_1),$$

and conjecture an isomorphism

$$\mathcal{B}_3(K)\otimes\mathbf{Q} \xrightarrow{\sim} K_5(K)\otimes\mathbf{Q} = K_5(\mathcal{O}_K)\otimes\mathbf{Q},$$

such that composition with Borel's regulator map is given by the map $\rho_{3,K}\otimes\mathbf{Q}$, where

$$\rho_{3,K} : \mathcal{B}_3(K) \longrightarrow \mathbf{R}^{r_1+r_2}$$

is given by

$$\rho_{3,K}\left(\sum_i n_i[x_i]\right) = \left(\sum_i n_i D_3(\sigma(x_i))\right)_\sigma,$$

where the σ run over both the real and complex embeddings of K. The real embeddings must be taken into account because D_3 does not vanish on real arguments. For the zeta-function $\zeta_K(s)$ of K one has Goncharov's result, cf. [Gon]:

Theorem 5.3.2 *Let* $\Delta : Z(K) \longrightarrow \wedge^2 K^* \otimes K^*$ *be given by*

$$\Delta\left(\sum_i n_i[x_i]\right) = \sum_i n_i\, x_i \wedge (1 - x_i) \otimes x_i \,.$$

Then there exist elements $y_1, \ldots, y_{r_1+r_2} \in \mathrm{Ker}(\Delta) \subset Z(K)$, *such that*

$$\zeta_K(s)_{s=3} \approx_{\mathbf{Q}^*} \frac{\pi^{3r_2}}{\sqrt{|D(K)|}} \det\left(D_3(\sigma(y_i))\right)_{\sigma,i} ,$$

where the σ *denote the real and (pairs of conjugate) complex embeddings of* K.

To conjecture about higher K-groups and values of $\zeta_K(s)_{s=m}$, $m = 3, \ldots$, one would like to try higher polylogarithms with suitable ('clean') functional equations. In [Za2] D. Zagier gives such functions, which he denotes by P_m. Actually D. Ramakrishnan [Ra1], in an abstract way, gave higher polylogarithms D_m which should play an important role in regulator problems and which generalized the Bloch-Wigner dilogarithm D_2. Zagier calculated these D_m explicitly and modified them a little bit to obtain even 'cleaner' expressions. Zagier's final higher polylogarithms are given by

$$P_m(x) = \Re_m\left(\sum_{j=0}^{m} \frac{2^j B_j}{j!} (\log|x|)^j Li_{m-j}(x)\right) .$$

Here we use $\Re_m$ for the real part $\Re$ if m is odd, and the imaginary part $\Im$ if m is even, and B_j for the Bernoulli numbers. The P_m are one-valued, real analytic on $\mathbf{P}^1_{\mathbf{C}} \setminus \{0, 1, \infty\}$ and continuous on all of $\mathbf{P}^1_{\mathbf{C}}$. Furthermore they are $(-1)^{m-1}$-symmetric with respect to $x \mapsto \dfrac{1}{x}$ and $x \mapsto \bar{x}$. Concerning functional equations for P_m, Zagier [Za2] proves

Proposition 5.3.1 *Let* $\{n_i, x_i(t)\}$ *be a collection of integers* n_i *and rational functions of one variable* $x_i(t)$ *satisfying the identity*

$$\sum_i n_i [x_i(t)]^{m-2} \otimes ([x_i(t)] \wedge [1 - x_i(t)]) = 0$$

in $(\mathrm{Sym}^{m-2}(\mathbf{C}(t)^*) \otimes \wedge^2(\mathbf{C}(t)^*)) \otimes_{\mathbf{Z}} \mathbf{Q}$. *Then*

$$\sum_i n_i P_m(x_i(t)) = \mathit{constant}\,.$$

Next, for any $\alpha \in \mathbf{P}^1_K$ let

$$\phi_\alpha : Z(\mathbf{Q}(t)) \longrightarrow Z(K)$$

be the evaluation map defined on generators by $[x(t)] \mapsto [x(\alpha)]$. Write

$$\mathcal{A}'_m(K) = \operatorname{Ker}\left\{Z(K) \xrightarrow{\beta_m} \operatorname{Sym}^{m-2}(K^* \otimes \mathbf{Q}) \otimes \Lambda^2(K^* \otimes \mathbf{Q})\right\},$$

where β_m is given on generators by

$$\beta_m([x]) = x^{m-2} \otimes (x \wedge (1-x)), \ \ x \in \mathbf{P}^1_K \setminus \{0,1,\infty\},$$

and $\beta_m([x]) = 0$ for $x \in \{0,1,\infty\}$. Let $\mathcal{C}_m(K)$ be the subgroup of $\mathcal{A}'_m(K)$ generated by all images $(\phi_\alpha - \phi_\beta)(\mathcal{A}'_m(\mathbf{Q}(t)))$ as α and β range over $\mathbf{P}^1_K$. Finally, set

$\mathcal{A}_m(K) =$

$$\operatorname{Ker}\left\{\mathcal{A}'_m(K) \xrightarrow{\imath} K^* \otimes \mathcal{A}'_{m-1}(K) \xrightarrow{Id\otimes\pi_{m-1}} K^* \otimes Z(K)/\mathcal{C}_{m-1}(K)\right\},$$

where $\imath : Z(K) \longrightarrow K^* \otimes Z(K)$ is defined on generators by $\imath[x] = x \otimes [x]$. After all these preparations we can define the **higher Bloch group** $\mathcal{B}_m(K)$ as follows:

Definition 5.3.3 *The* m^{th} Bloch group $\mathcal{B}_m(K)$ *of* K *is defined as*

$$\mathcal{B}_m(K) = \mathcal{A}_m(K)/\mathcal{C}_m(K).$$

The group $\mathcal{C}_m(K)$ should be considered as the group spanned by the functional equations of the P_m. Again, as in the previous cases, there are maps

$$\rho_{m,K} : \mathcal{B}_m(K) \longrightarrow \mathbf{R}^{r_2} \text{ or } \mathbf{R}^{r_1+r_2},$$

according to the parity of m. We can state Zagier's Main Conjecture, cf. [Za2]:

Conjecture 5.3.1 *There is a canonical map*

$$\mathcal{B}_m(K) \longrightarrow K_{2m-1}(K)$$

with finite kernel and cokernel, whose composite with the Borel regulator mapping coincides with $\rho_{m,K}$.

For the zeta-function $\zeta_K(s)$ one may conjecture results analogous to Zagier's or Goncharov's mentioned above.[2] In [Za2] the reader will find lots of interesting numerical examples.

To finish this section, we mention some other aspects of the dilogarithm D_2 and, presumebly, of Ramakrishnan's or Zagier's higher polylogarithms, cf. [Ra3]. An important property of D_2 in relation to the regulator map is that D_2 determines a non-trivial class in continuous cohomology $H^3_{cont}(\mathrm{GL}_2(\mathbf{C}), \mathbf{R})$ by sending (g_1, g_2, g_3, g_4) to the real number

$$D_2(\{g_1(\infty), g_2(\infty), g_3(\infty), g_4(\infty)\}),$$

where $\{..,..,..,..\}$ denotes the cross ratio. Forgetting the topology one obtains a non-trivial class

$$\bar{D}_2 \in H^3(\mathrm{GL}_2(\mathbf{C}), \mathbf{R}) = \mathrm{Hom}(H_3(\mathrm{GL}_2(\mathbf{C}), \mathbf{Z}), \mathbf{R}).$$

Combining with the Hurewicz map this leads to the map

$$K_3(\mathrm{C}) = \pi_3(\mathrm{B_{\bullet}GL}(\mathrm{C})^+) \longrightarrow H_3(\mathrm{GL}(\mathrm{C}), \mathbf{Z}) \xrightarrow{2\pi i \bar{D}_2} \mathbf{R}(1).$$

This may be used to calculate explicitly the regulator (map) by means of the formulas of the previous section.

For the higher K-groups of $\mathbf{C}$, Ramakrishnan was led to ask (cf. [Ra3]): For every $m \geq 1$, does there exist, for some $n \geq m$, a function

$$f_{m,n} : (\mathrm{GL}_n(\mathbf{C}))^{2m} \longrightarrow \mathbf{P}^1_{\mathbf{C}},$$

invariant under the diagonal action of $\mathrm{GL}_n(\mathbf{C})$, and factoring through a product of Grassmannians, such that $\beta_m = D_m \circ f_{m,n}$ gives rise to a non-trivial, indecomposable stable class in $H^{2m-1}_{cont}(\mathrm{GL}_n(\mathbf{C}), \mathbf{R})$? Here D_m is a generalization of D_2 based on Li_m. We know now that there are candidates for the D_m. It remains to verify their geometric properties. One also hopes that the D_m will play a role in the definition of the higher complexes $\mathbf{Z}(m)^{\bullet}_{\mathcal{M}}$.

---

[2] It seems that Beilinson has made substantial progress on Zagier's Main Conjecture.

In the case of an affirmative answer to the above question, one could realize concretely Borel's regulator maps

$$K_{2m-1}(\mathbf{C}) \xrightarrow{\text{Hurewicz}} H_{2m-1}(B_{\bullet}GL(\mathbf{C}), \mathbf{Z}) \xrightarrow{(2\pi i)^{m-1}\beta} \mathbf{R}(m-1),$$

and then also, by the procedure of the previous section, $r_{m,Borel}$ for a number field K.

5.4 Riemann Surfaces

The second example concerns the global regulator map of a Riemann surface. It is due mainly to Beilinson and Bloch.

Let X be a smooth, projective curve over $\mathbf{Q}$ and let $X(\mathbf{C})$ be the corresponding compact Riemann surface. Let S be a finite subset of $X(\mathbf{C})$ and write $Y_S(\mathbf{C})$ for $X(\mathbf{C})\backslash S$. Then there is the long exact sequence

$$0 \longrightarrow H^1(X(\mathbf{C}), \mathbf{C}^*) \longrightarrow H^1(Y_S(\mathbf{C}), \mathbf{C}^*) \xrightarrow{\coprod_{x\in S}\partial_x} \coprod_{x\in S} \mathbf{C}^* \longrightarrow$$

$$\longrightarrow H^2(X(\mathbf{C}), \mathbf{C}^*) \longrightarrow \dots \tag{5.7}$$

In this sequence $H^1(Y_S(\mathbf{C}), \mathbf{C}^*) = \mathbf{H}^1(X(\mathbf{C}), \mathcal{O}_X^*(\star S) \xrightarrow{\text{dlog}} \Omega_X^1(\log S))$, where $\mathcal{O}_X^*(\star S)$ is the sheaf of functions, holomorphic and invertible on Y_S and $\Omega_X^1(\log S)$ is the sheaf of 1-forms, holomorphic on Y_S and with at most first order poles at S. The exact sequence (5.4) follows from the distinguished triangle

$$\begin{array}{ccc} & \left(\coprod_{x\in S} \mathbf{Z}_x \hookrightarrow \coprod_{x\in S} \mathbf{C}_x\right) \xrightarrow{\sim} \coprod_{x\in S} \mathbf{C}_x^*[-1] & \\ {}^{[1]}\swarrow & & \nwarrow^{(\text{ord},\,\text{res})} \\ \mathbf{C}_X^* \xrightarrow{\sim} (\mathcal{O}_X^* \xrightarrow{d\log} \Omega_X^1) & \longrightarrow & (\mathcal{O}_X^*(\star S) \xrightarrow{d\log} \Omega_X^1(\log S)) \end{array} \tag{5.8}$$

where $\xrightarrow{\sim}$ means quasi-isomorphism, $\text{ord}_x : \mathcal{O}_X^*(\star S) \longrightarrow \mathbf{Z}_x$ is the order at x and $\text{res}_x : \Omega_X^1(\log S) \longrightarrow \mathbf{C}_x$ is the Poincaré residue map.

They are related by $\operatorname{res}_x d\log = \operatorname{ord}_x$. The quasi-isomorphism at the top of the triangle follows from the short exact sequence

$$0 \longrightarrow \mathbf{Z} \longrightarrow \mathbf{C} \xrightarrow{e} \mathbf{C}^* \longrightarrow 0,$$

with $e(x) = \exp(2\pi i x)$. Then for an element $\alpha \in H^1(Y_S(\mathbf{C}), \mathbf{C}^*)$ represented by the differential form $\omega \in \Omega^1_X(\log S)$ one has : $\partial_x(\alpha) = e(\operatorname{res}_x \omega)$. Taking the limit over all finite subsets $S \subset X(\mathbf{C})$ one obtains the sequence

$$H^1(X(\mathbf{C}), \mathbf{C}^*) \xrightarrow{\mu} \lim_{\substack{S \subset X(\mathbf{C}) \\ finite}} H^1(Y_S(\mathbf{C}), \mathbf{C}^*) \xrightarrow{\coprod_{x \in X(\mathbf{C})} \partial_x} \coprod_{x \in X(\mathbf{C})} \mathbf{C}^*, \tag{5.9}$$

where μ is injective.

This sequence will be related to an exact sequence in algebraic K-theory. To see how this is done, let f and g be meromorphic functions on $X(\mathbf{C})$. Then the function $\dfrac{f^{\operatorname{ord}_x g}}{g^{\operatorname{ord}_x f}}$ has order zero at any $x \in X(\mathbf{C})$, so its value at every x is well defined.

Definition 5.4.1 *The* tame symbol *of f and g at $x \in X(\mathbf{C})$ is the expression*

$$T_x(f,g) = (-1)^{\operatorname{ord}_x f \operatorname{ord}_x g} \cdot \frac{f^{\operatorname{ord}_x g}}{g^{\operatorname{ord}_x f}}(x) \in \mathbf{C}^*.$$

This tame symbol has the following properties:

(*i*) $T_x(f\,g, h) = T_x(f,h)\,T_x(g,h)$ and analogously for $T_x(f, g\,h)$;

(*ii*) $T_x(f,g)\,T_x(g,f) = 1$;

(*iii*) $T_x(1-f, f) = 1$.

Writing $\mathbf{C}(X)$ for the field of meromorphic functions on $X(\mathbf{C})$, one has the so-called localization exact sequence:

$$K_2(X(\mathbf{C})) \xrightarrow{\lambda} K_2(\mathbf{C}(X)) \xrightarrow{\coprod_{x \in X(\mathbf{C})} T_x} \coprod_{x \in X(\mathbf{C})} \mathbf{C}^*. \tag{5.10}$$

Here we have used Matsumoto's theorem 4.4.1 which says that

$$K_2(\mathbf{C}(X)) = \mathbf{C}(X)^* \otimes \mathbf{C}(X)^* / \langle f \otimes (1-f) \mid f \in \mathbf{C}(X)^* \setminus \{0,1\} \rangle .$$

Also, λ is injective modulo torsion.

The searched for commutative diagram has the form:

$$\begin{array}{ccccc}
K_2(X(\mathbf{C})) & \xrightarrow{\lambda} & K_2(\mathbf{C}(X)) & \xrightarrow{\coprod_{x\in X(\mathbf{C})} T_x} & \coprod_{x\in X(\mathbf{C})} \mathbf{C}^* \\
\downarrow r_1 & & \downarrow r_0 & & \| \\
H^1(X(\mathbf{C}),\mathbf{C}^*) & \xrightarrow{\mu} & \lim\limits_{\substack{S\subset X(\mathbf{C}) \\ finite}} H^1(Y_S(\mathbf{C}),\mathbf{C}^*) & \xrightarrow{\coprod_{x\in X(\mathbf{C})} \partial_x} & \coprod_{x\in X(\mathbf{C})} \mathbf{C}^*
\end{array} \tag{5.11}$$

We are left with the construction of r_0, which will give the desired **regulator map**

$$r_1 : K_2(X(\mathbf{C})) \longrightarrow H^1(X(\mathbf{C}),\mathbf{C}^*) \simeq H^2_{\mathcal{D}}(X,\mathbf{Z}(2))$$

because the Deligne complex $\mathbf{Z}(2)_{\mathcal{D}}$ is quasi-isomorphic with the complex $(\mathcal{O}^*_X \xrightarrow{d\log} \Omega^1_X)[-1]$ by the following commutative diagram:

$$\begin{array}{ccccccc}
\mathbf{Z}(2)_{\mathcal{D}} : 0 \longrightarrow \mathbf{Z}(2) & \longrightarrow & \mathcal{O}_X & \longrightarrow & \Omega^1_X & \longrightarrow & 0 \\
 & & \downarrow \exp\frac{x}{2\pi i} & & \downarrow -(2\pi i)^{-1} & & \\
 & & \mathcal{O}^*_X & \xrightarrow{d\log} & \Omega^1_X & \longrightarrow & 0
\end{array} \tag{5.12}$$

The proof of the commutativity of the diagram (5.11) follows from an explicit local calculation once r_0 is known.

To construct r_0, let $f,g \in \mathbf{C}(X)^*$ and write $S_{f,g} \subset X(\mathbf{C})$ for the finite set of zeroes and poles of f and g. Thus $S_{f,g} = \mathrm{div}(f) \cup \mathrm{div}(g)$. We define the map

$$r_0 : \mathcal{O}^*_{Y_{S_{f,g}}(\mathbf{C})} \otimes \mathcal{O}^*_{Y_{S_{f,g}}(\mathbf{C})} \longrightarrow H^1(Y_{S_{f,g}}(\mathbf{C}),\mathbf{C}^*) \tag{5.13}$$

where $H^1(Y_{S_{f,g}}(\mathbf{C}),\mathbf{C}^*)$ is the group of line bundles on $Y_{S_{f,g}}(\mathbf{C})$ with flat connection, by the following procedure. First, to abbreviate notation, write $Y_{f,g}$ for $Y_{S_{f,g}}(\mathbf{C})\ldots$ etc. The pair (f,g) defines a map

$$(f,g) : Y_{f,g} \longrightarrow \mathbf{C}^* \times \mathbf{C}^*,$$

but on $\mathbf{C}^* \times \mathbf{C}^*$ one has a universal (Heisenberg) line bundle with connection $(\mathcal{L},\nabla)$, (cf. [Ra3]). One defines $r_0(f,g) = (f,g)^*(\mathcal{L},\nabla)$. To show that this is well defined on $K_2(\mathbf{C}(Y_{f,g}))$ one needs the following facts:

(*i*) $r_0(f.g,h) = r_0(f,h) \otimes r_0(g,h)$ on $Y_{f.g,h}$;

(*ii*) $r_0(f,g) \otimes r_0(g,f) = (\mathcal{O}_{Y_{f,g}}, d)$, the trivial bundle with trivial connection, given by the 1-form

$$\omega = \frac{-1}{2\pi i}\Big(\log(f)\,\frac{dg}{g} + \log(g)\,\frac{df}{f}\Big) = \frac{-1}{2\pi i}\, d(\log(f)\,\log(g))\,,$$

thus $e(\mathrm{res}_x\omega) = 1 \in \mathbf{C}^*$.

(*iii*) $r_0(1-f,f) = (\mathcal{O}_{Y_{1-f,f}}, d)$, the trivial bundle ...etc.

These facts follow from an explicit description of $r_0(f,g)$ as a 1-form on $Y_{f,g}$. Observing that the map r_0 also corresponds to the cup product

$$\cup :\ H^1_{\mathcal{D}}(Y_{f,g}, \mathbf{Z}(1)) \otimes H^1_{\mathcal{D}}(Y_{f,g}, \mathbf{Z}(1)) \longrightarrow H^2_{\mathcal{D}}(Y_{f,g}, \mathbf{Z}(2))\,,$$

coming from the pairing $\mathbf{Z}(1)_{\mathcal{D}} \otimes \mathbf{Z}(1)_{\mathcal{D}} \longrightarrow \mathbf{Z}(2)_{\mathcal{D}}$ as described in Chapter 3, one sees that the differential form associated to $r_0(f,g)$ is

$$-(2\pi i)^{-1}\log(f)\, d\log(g)$$

and therefore

$$\partial_x(r_0(f,g)) = \exp\left(\frac{-1}{2\pi i}\oint_{\partial D}\log(f)\,\frac{dg}{g}\right),$$

where D is a small neighbourhood of x.

The map $r_1 : K_2(X(\mathbf{C})) \longrightarrow H^1(X(\mathbf{C}), \mathbf{C}^*)$ is sometimes combined with $2\pi i \otimes \log|.| : \mathbf{C}^* \longrightarrow \mathbf{R}(1)$ to give the actual **regulator map**

$$r :\ K_2(X(\mathbf{C})) \longrightarrow H^1(X(\mathbf{C}), \mathbf{R}(1))$$

and, in fact, one can prove that for a smooth, projective curve X defined over $\mathbf{Q}$, and denoting by the superscript $^+$ the invariants under de Rham conjugation, one has

$$r :\ K_2(X) \longrightarrow H^1(X(\mathbf{C}), \mathbf{R}(1))^+ = H^2_{\mathcal{D}}(X_{/\mathbf{R}}, \mathbf{R}(2))\,. \qquad (5.14)$$

Let $X_{\mathbf{Z}}$ be a regular model of $X/\mathbf{Q}$ over $\mathbf{Z}$. This exists by a result of Abhyankar, and it induces a natural map from $K_2(X_{\mathbf{Z}})$ to $K_2(X)$.

The image of this map is independent of the model $X_{\mathbf{Z}}$. With this notion of regular model, Beilinson and Bloch formulated the following conjecture:

Conjecture 5.4.1 (Beilinson-Bloch) *Let X over $\mathbf{Q}$ be a smooth, projective curve of genus g, and let*

$$L(X/\mathbf{Q}, s) = c_g\, s^g(1 + \ldots)$$

be the L-function of $X/\mathbf{Q}$ at $s = 0$. Let $X_{\mathbf{Z}}$ be a regular model of X over $\mathbf{Z}$ and write r also for the composition of the natural map $K_2(X_{\mathbf{Z}}) \longrightarrow K_2(X)$ and $r : K_2(X) \longrightarrow H^2_{\mathcal{D}}(X_{/\mathbf{R}}, \mathbf{R}(2))$. Then:

(i) $r(K_2(X_{\mathbf{Z}})) \otimes \mathbf{Q}$ defines a $\mathbf{Q}$-structure on $H^2_{\mathcal{D}}(X_{/\mathbf{R}}, \mathbf{R}(2))$;

(ii) $R = \det(r) \approx_{\mathbf{Q}^\bullet} c_g$;

(iii) $\mathrm{Ker}(r) \subset K_2(X_{\mathbf{Z}})$ is finite.

Remark 5.4.1 The fact that $\mathrm{ord}_{s=0}\, L(X/\mathbf{Q}, s) = g$ would follow from the Standard Conjectures on L-functions.

Remark 5.4.2 Originally the conjecture was formulated without introducing the regular model $X_{\mathbf{Z}}$ of $X/\mathbf{Q}$. However, in the fall of 1981, Bloch and Grayson (cf. [BG]) made computer calculations for several elliptic curves $X/\mathbf{Q}$. These computations revealed that the regulator map

$$r \otimes \mathbf{R} : H^2_{\mathcal{M}}(X, \mathbf{Q}(2)) \otimes \mathbf{R} \longrightarrow H^2_{\mathcal{D}}(X_{/\mathbf{R}}, \mathbf{R}(2))$$

is not always injective, or in other words, the rank of $K_2(X)$ could be larger than one. This contradicted another conjecture of Bloch which had been verified in many cases. This conjecture of Bloch concerns the value of $L(X/\mathbf{Q}, s)$ at $s = 2$. Bloch and Grayson suggested to look at the Néron model $\mathcal{E} = X_{\mathbf{Z}}$ of $X/\mathbf{Q}$ and then hoped that $K_2(X_{\mathbf{Z}})$ would have rank one. The localization exact sequence

$$K_2(X_{\mathbf{Z}}) \longrightarrow K_2(X) \longrightarrow \coprod_p K_1'(X_{\mathbf{F}_p})$$

and the fact that for primes p where X has good reduction, $K_1'(X_{\mathbf{F}_p})$ is torsion, implies that a bad fibre $X_{\mathbf{F}_p}$ with $K_1'(X_{\mathbf{F}_p}) = \mathbf{Z} \oplus \mathit{tors}$. causes the rank of $K_2(X)$ to be larger than one. Actually, this can happen only if $X_{\mathbf{F}_p}$ is of Kodaira type I_ν, $\nu \geq 1$.

5.5 Models over Spec(Z)

In the sequel we will always assume that our varieties $X/\mathbf{Q}$ admit a regular model $X_{\mathbf{Z}}$ over Spec($\mathbf{Z}$), i.e. there exists a regular proper scheme $X_{\mathbf{Z}}$ over Spec($\mathbf{Z}$) such that $X_{\mathbf{Z}} \otimes \mathbf{Q} \cong X$. This may be compared to the case when X is an abelian variety over $\mathbf{Q}$ and the regular model is the Néron model.

Definition 5.5.1

$$\begin{aligned} H^{\bullet}_{\mathcal{M}}(X, \mathbf{Q}(\star))_{\mathbf{Z}} &= \operatorname{Im}\big(H^{\bullet}_{\mathcal{M}}(X_{\mathbf{Z}}, \mathbf{Q}(\star)) \longrightarrow H^{\bullet}_{\mathcal{M}}(X, \mathbf{Q}(\star))\big) \subseteq \\ &\subseteq H^{\bullet}_{\mathcal{M}}(X, \mathbf{Q}(\star)) \textit{ as a subring}. \end{aligned}$$

One can show that this definition is independent of the choice of the regular model, provided that it exists.

Anyhow, it is possible to find a **proper, flat** model $X_{\mathbf{Z}}$ of $X/\mathbf{Q}$. The following K-theoretic conjecture can be stated (cf. [Sc]):

Conjecture 5.5.1 *For a proper, flat model $X_{\mathbf{Z}}$ of $X/\mathbf{Q}$ the image*

$$\operatorname{Im}\big(K'_{*}(X_{\mathbf{Z}}) \otimes \mathbf{Q} \longrightarrow K_{*}(X) \otimes \mathbf{Q}\big)$$

is independent of the choice of $X_{\mathbf{Z}}$ and is compatible with the Adams operations and the formation of inverse images with respect to x.

If one does not want to assume the existence of a regular model, but is willing to accept the above conjecture, then one may alter the definition of $H^{\bullet}_{\mathcal{M}}(X, \mathbf{Q}(\star))_{\mathbf{Z}}$ and restate it as follows:

Definition 5.5.2

$$H^{i}_{\mathcal{M}}(X, \mathbf{Q}(j))_{\mathbf{Z}} =$$
$$= \operatorname{Im}\big(K'_{2j-i}(X_{\mathbf{Z}}) \otimes \mathbf{Q} \longrightarrow K_{2j-i}(X) \otimes \mathbf{Q}\big) \subseteq H^{i}_{\mathcal{M}}(X, \mathbf{Q}(j)).$$

Anyhow, the following conjecture remains:

Conjecture 5.5.2 (*i*) $H^{i}_{\mathcal{M}}(X, \mathbf{Q}(j))_{\mathbf{Z}} = H^{i}_{\mathcal{M}}(X, \mathbf{Q}(j))$ *for any* (i, j) *that do not satisfy* $i \leq j \leq 2i - 1$ *and* $i \leq \dim(X) + 1$.

(*ii*) $H^{i}_{\mathcal{M}}(X, \mathbf{Q}(j))/H^{i}_{\mathcal{M}}(X, \mathbf{Q}(j))_{\mathbf{Z}}$, $i \leq 2j - 2$, *depends only on the bad fibres of* $X_{\mathbf{Z}}$.

Remark 5.5.1 G. Harder proved that this conjecture holds for curves.

5.6 Deligne's Conjecture

The third example to motivate the formulation of the first Beilinson Conjecture is a conjecture of P. Deligne (cf. [De2]). To state it, let X be a smooth, projective variety defined over $\mathbf{Q}$.

Definition 5.6.1 *The point $s = m$ is called* critical *if $L_\infty(X, s)$ does not have a pole at either $s = m$ or $s = i + 1 - m$.*

Here L_∞ is the factor at infinity of the L-function $L(X, s)$ of $X/\mathbf{Q}$, and i is the order of the ℓ-adic cohomology of X defining this L-function. For details we refer to Chapter 3.

This means that $L(X, m) \neq 0$ or, equivalently, that the dimension of the Deligne cohomology $\dim\, H_{\mathcal{D}}^{i+1}\,(X_{/\mathbf{R}}, \mathbf{R}(i + 1 - m)) = 0$, for $m < \frac{i}{2}$. One obtains the immediate

Corollary 5.6.1 *$s = m$ is critical if and only if*

$$F^{i+1-m} H^i_{DR}(X_{/\mathbf{R}}) \xrightarrow[\sim]{I^\vee} H^i(X_{/\mathbf{R}}, \mathbf{R}(i - m))\,.$$

Here both sides admit $\mathbf{Q}$–structures: the left hand side the one coming from the algebraic de Rham cohomology, and the right hand side the one coming from $\mathbf{Q}$–coefficients. Comparing both $\mathbf{Q}$–structures gives Beilinson's period

$$c_{Beilinson}(m) = \det(I^\vee) \in \mathbf{R}^*/\mathbf{Q}^*\,.$$

However, Deligne's approach was different and formulated in the language of motives, thus combining Betti, de Rham and ℓ-adic cohomology as the various realizations of the underlying motive. For a more detailed account of Deligne motives[3] we refer to Chapter 8. For the moment it suffices to know that a smooth, projective variety over a field defines a motive and that the motive combines in some sense the various cohomologies attached to the variety.

The advantage of this approach is that one can freely use suitable Tate twists, Poincaré duality and a hard Lefschetz theorem to dualize the motive M such that if M comes from a smooth, projective variety X

[3]i.e. motives for absolute Hodge cycles in the language of Chapter 8.

and can be written as $M = H^i(X)(m)$, then one has for the dual motive $M^\vee = H^i(X)(i-m)$. Actually, the category of motives constructed in this way is much nicer than the category of smooth, projective varieties, e.g. it is abelian and has a natural tensor product etc.

For Deligne's conjecture explicit use is made only of the Betti and the de Rham realizations, $H_B(M)$ and $H_{DR}(M)$, respectively, of the motive $M = H^i(X)(m)$. The ℓ-adic realizations come in via the L-function. To abbreviate we omit the M in $H_B(M)$ and $H_{DR}(M)$. Then H_B is a $\mathbf{Q}$–vector space with Hodge decomposition of the associated $H_B \otimes \mathbf{C}$ and an action of F_∞, the non-trivial element of $\mathrm{Gal}(\mathbf{C}/\mathbf{R})$. For the Tate twist $H_B(n)$ of H_B one may observe that it is the $\mathbf{Q}$–subspace $H_B \otimes (2\pi i)^n\, \mathbf{Q}$ of $H_B \otimes \mathbf{C}$. The action of F_∞ on $H_B(n)$ is given by

$$F_\infty | H_B(n) = (-1)^n\, F_\infty | H_B\,.$$

The de Rham realization H_{DR} is a $\mathbf{Q}$–vector space with decreasing Hodge filtration $F^\bullet H_{DR}$. Furthermore, for the Tate twists one has $H_{DR}(n) = H_{DR}$ and $F^p H_{DR}(n) = F^{p+n} H_{DR}$. If M is defined over $\mathbf{R}$ one has an isomorphism

$$H_B \otimes \mathbf{C} \xrightarrow[\sim]{I_{DR}} H_{DR} \otimes \mathbf{C}\,, \tag{5.15}$$

identifying H_{DR} as the subspace $(H_B \otimes \mathbf{C})^{DR}$ of $H_B \otimes \mathbf{C}$.

For a smooth, projective variety $X/\mathbf{Q}$ the motive $M = H^i(X)(m)$ is pure of weight $i - 2m$ and in case m is critical F_∞ acts as ± 1 on $H^{p,p}$, where $H^{p,p}$ is the (p,p)–part of the Hodge decomposition $H_B \otimes \mathbf{C} = \bigoplus_{i,j} H^{i,j}$, if the weight equals $2p$. One can decompose H_B into ± 1-eigenspaces of F_∞: $H_B = H_B^+ \oplus H_B^-$ and I_{DR} gives filtration steps

$$(F^\pm H_{DR}) \otimes \mathbf{C} = I_{DR}\Big(\bigoplus_{p>q} H^{p,q} \bigoplus (H^{p,p})^\pm\Big)\,.$$

Then write $H_{DR}^\pm = H_{DR}/F^\mp H_{DR}$. Now $F_\infty H^{p,q} = H^{q,p}$ and one obtains isomorphisms:

$$I^\pm : H_B^\pm \otimes \mathbf{C} \longrightarrow H_B \otimes \mathbf{C} \xrightarrow[\sim]{I_{DR}} H_{DR} \otimes \mathbf{C} \longrightarrow H_{DR}^\pm \otimes \mathbf{C}\,. \tag{5.16}$$

Both extremes of this sequence have $\mathbf{Q}$–structures, so one may define the **Deligne periods**

$$c^{\pm}_{Deligne}(M) = \det(I^{\pm})$$

with respect to $\mathbf{Q}$–bases of $H^{\pm}_B$ and $H^{\pm}_{DR}$. Actually, I^+ is defined over $\mathbf{R}$, i.e.

$$I^+ : H^+_B \otimes \mathbf{R} \longrightarrow H^+_{DR} \otimes \mathbf{R}, \text{ so } c^+_{Deligne}(M) \in \mathbf{R}^*/\mathbf{Q}^*.$$

Using the ℓ-adic realizations of M one can define the L-function $L(M,s)$ of the motive M. For $M = H^i(X)(m)$ one has $L(M,0) = L(X,m)$.

Conjecture 5.6.1 (Deligne) *One has:*

$$L(M,0) \approx_{\mathbf{Q}^*} c^+_{Deligne}(M)$$

or, writing $c^+_{Deligne}(m)$ for $c^+_{Deligne}(M)$ when m is critical, one has:

$$L(X,s)_{s=m} \approx_{\mathbf{Q}^*} c^+_{Deligne}(m).$$

Example 5.6.1 $X = \mathrm{Spec}(\mathbf{Q})$, $L(X,s) = \zeta(s)$, Riemann's ζ-function. Then $s = m$ is critical iff $m > 0$ and even or $m < 0$ and odd, and

$$c^+(m) = \begin{cases} (2\pi i)^m & , \ m > 0 \text{ and even} \\ 1 & , \ m < 0 \text{ and odd.} \end{cases}$$

Remark 5.6.1 Let $X = \mathrm{Spec}(K)$, K a totally real number field. Then $s = 1 - 2m$, $m \geq 1$, is critical and $c^+_{Deligne}(1-2m) = 1$. Deligne's conjecture is true by a result of Siegel:

$$\zeta_K(1-2m) \in \mathbf{Q}^*, \ m \geq 1.$$

Remark 5.6.2 Let $X = E$ be an elliptic curve defined over $\mathbf{Q}$ with $r = \mathrm{rank}(E(\mathbf{Q})) = 0$. Then $m = 1$ is a critical point and the motive $M = H^1(E)(1)$ is isomorphic with the dual $H_1(E)$ of $H^1(E)$. One has $L(M,0) = L(E,1)$. Also, $F^+H^1_{DR}(E) = H^0(E,\Omega^1_E)$ and $H^0(E,\Omega^1_E)$ has generator ω defined over $\mathbf{Q}$. Let c be a generator of $H_1(E_{/\mathbf{R}},\mathbf{Q})$, then

$$c^+_{Deligne}(1) = \int_c \omega = \int_{E(\mathbf{R})} |\omega|$$

and one sees that Deligne's conjecture is compatible with the second Birch & Swinnerton-Dyer Conjecture. This is also true, *mutatis mutandis*, for higher dimensional abelian varieties defined over $\mathbf{Q}$ (or, more generally, any number field).

Remark 5.6.3 Deligne proved his conjecture for Artin motives, i.e. finite products of number fields, where a morphism between two such products is given by a correspondence defined over $\mathbf{Q}$.[4].

The relation between Beilinson's and Deligne's periods is given by the following theorem.

Theorem 5.6.1 *Let $X/\mathbf{Q}$ be a smooth, projective variety, and let m be a critical point, then*

$$c_{Beilinson}(m) = c^{+}_{Deligne}(m)\,.$$

For the proof one transposes the map I^{+} and uses Poincaré duality to get a map which coincides with $I^{\vee}$, (cf. [HW]).

5.7 Beilinson's First Conjecture

We are now in a position to give an accurate statement of Beilinson's first conjecture for smooth, projective varieties defined over the rational numbers (or any number field).

So let X be such a variety and let $L(X,s)$ be the L-function of X defined by the $\mathrm{Gal}(\bar{\mathbf{Q}}/\mathbf{Q})$-modules $H^i(X\otimes\bar{\mathbf{Q}},\mathbf{Q}_\ell)\cong H^i(X_p\otimes\bar{\mathbf{F}}_p,\mathbf{Q}_\ell)$, $\ell\neq p$, for fixed i, $0\leq i\leq 2\dim(X)$, (cf. Chapter 3).

Recall that both motivic cohomology $H^{\bullet}_{\mathcal{M}}(X,\mathbf{Q}(\star))$ and Deligne-Beilinson cohomology $H^{\bullet}_{\mathcal{D}}(X,\mathbf{R}(\star))$ are Poincaré duality theories, so admit the construction of Chern classes and Chern characters as described in Chapter 4. According to Beilinson's general belief in the universality of motivic cohomology, the formalism of Chern classes and Chern characters leads to a natural transformation

$$r:\ H^{\bullet}_{\mathcal{M}}(X,\mathbf{Q}(\star))\longrightarrow H^{\bullet}_{\mathcal{D}}(X,\mathbf{R}(\star))\,.$$

[4]Cf. Chapter 10 for Deligne's Conjecture for Hecke L-functions

Combining with $H^{\bullet}_{\mathcal{M}}(X, \mathbf{Q}(\star)) \longrightarrow H^{\bullet}_{\mathcal{M}}(X_{/\mathbf{R}}, \mathbf{Q}(\star))$ and taking de Rham invariants finally gives **Beilinson's regulator map**

$$r_{\mathcal{D}} : H^{\bullet}_{\mathcal{M}}(X, \mathbf{Q}(\star)) \longrightarrow H^{\bullet}_{\mathcal{D}}(X_{/\mathbf{R}}, \mathbf{R}(\star)) . \tag{5.17}$$

The latter spaces $H^{\bullet}_{\mathcal{D}}$ have a natural $\mathbf{Q}$–structure, so any other $\mathbf{Q}$–structure on $H^{\bullet}_{\mathcal{D}}$ has a well defined **volume**. Taking into account the foregoing examples, one can state Beilinson's first conjecture:

Conjecture 5.7.1 (Beilinson I) *For* $m < \frac{i}{2}$,
(i) $r_{\mathcal{D}} : H^{i+1}_{\mathcal{M}}(X, \mathbf{Q}(i+1-m))_{\mathbf{Z}} \longrightarrow H^{i+1}_{\mathcal{D}}(X_{/\mathbf{R}}, \mathbf{R}(i+1-m))$ *induces a* $\mathbf{Q}$*–structure on Deligne cohomology.*

(ii) The volume $R_{\mathcal{D}} = \det(r_{\mathcal{D}})$ *of this* $\mathbf{Q}$*-structure is equal (up to a non-zero rational number) to the first non-zero coefficient of the Taylor series expansion of* $L(X, s)$ *at* $s = m$*, in other words:*

$$R_{\mathcal{D}} \approx_{\mathbf{Q}^*} L^*(X, s)_{s=m} .$$

Remark 5.7.1 For $X = \operatorname{Spec}(K)$, K a number field, and $i = 0$, the conjecture is verified. This follows from Theorem 5.2.2.

Remark 5.7.2 For critical m, $H^{i+1}_{\mathcal{M}}(X, \mathbf{Q}(i+1-m))$ should be zero and the $\mathbf{Q}$–structure should coincide with the one described in Section 5.6 above. Thus for critical m, $m < \frac{i}{2}$, the conjecture reduces to Deligne's conjecture.

Remark 5.7.3 Let $E/\mathbf{Q}$ be an elliptic curve with (minimal) Néron model $\mathcal{E}$ over $\operatorname{Spec}(\mathbf{Z})$, and let $P = X \cup Y \subset E$ be the union of two finite sets of points $X = \{x_i\}$ and $Y = \{y_j\}$, such that $x_i - y_j \in E(\mathbf{Q})_{tors}$. A result of Bloch says that in this situation the space $H^2_{\mathcal{M}}(E \setminus P, \mathbf{Q}(2))$ decomposes into the direct sum of $H^2_{\mathcal{M}}(E, \mathbf{Q}(2))$ and a part generated by symbols of the form $\{f, a\} \otimes \mathbf{Q}$, where $f \in \mathcal{O}^*(E \setminus P)$ and $a \in \mathbf{Q}^*$. This implies in particular that there is a canonical projection

$$pr_{\mathcal{M}} : H^2_{\mathcal{M}}(E \setminus P, \mathbf{Q}(2)) \longrightarrow H^2_{\mathcal{M}}(E, \mathbf{Q}(2)) .$$

Let $f, g \in \mathcal{O}^*(E \setminus P)$ such that $\operatorname{div}(f) = \sum n_i x_i$ and $\operatorname{div}(g) = \sum m_j y_j$. They give rise to an element $\{f, g\} \in H^2_{\mathcal{M}}(E \setminus P, \mathbf{Q}(2)) \subset K_2(E \setminus P) \otimes \mathbf{Q}$, thus to an element $pr_{\mathcal{M}}\{f,g\} \in H^2_{\mathcal{M}}(E,\mathbf{Q}(2))$. This construction leads to a pairing, also denoted $\{\ ,\ \}$,

$$\{\ ,\ \} : \wedge^2 \mathbf{Q}[P]^0 \longrightarrow H^2_{\mathcal{M}}(E,\mathbf{Q}(2)),$$

defined by $\{\mathrm{div}(f),\mathrm{div}(g)\} = pr_{\mathcal{M}}\{f,g\}$. The superscript ^{0} means degree zero.

On the other hand, using example 3.2.8 , one can construct a commutative diagram

$$\begin{array}{ccc} \wedge^2 H^1_{\mathcal{D}}(E\backslash P_{/\mathbf{R}},\mathbf{R}(1)) & \xrightarrow{\ \cup\ } & H^2_{\mathcal{D}}(E\backslash P_{/\mathbf{R}},\mathbf{R}(2)) = H^1(E\backslash P,\mathbf{R}(1)) \\ \downarrow \mathrm{div} & & \downarrow pr_{\mathcal{D}} \\ \wedge^2 \mathbf{R}[P]^0 & \xrightarrow{\ [\ ,\ ]_{\mathcal{D}}\ } & H^2_{\mathcal{D}}(E_{/\mathbf{R}},\mathbf{R}(2)) = H^1(E,\mathbf{R}(1)) \\ \downarrow & & \downarrow \\ 0 & & 0 \end{array}$$

giving the pairing $[\ ,\ ]_{\mathcal{D}}$. In $H^1(E\backslash P,\mathbf{R}(1))$ and $H^1(E,\mathbf{R}(1))$, $E\backslash P$ and E are considered as analytic spaces over $\mathbf{R}$. Thus

$$H^1(E,\mathbf{R}(1)) \cong \mathrm{Hom}(F^1(E),\mathbf{R}), \quad \text{where} \quad F^1(E) = H^0(E(\mathbf{C}),\Omega^1_E)^{DR}.$$

Assume that E has **complex multiplication** , then one can show that $H^2_{\mathcal{M}}(E,\mathbf{Q}(2)) = H^2_{\mathcal{M}}(\mathcal{E},\mathbf{Q}(2))$, i.e. $H^2_{\mathcal{M}}(E,\mathbf{Q}(2)) = H^2_{\mathcal{M}}(E,\mathbf{Q}(2))_{\mathbf{Z}}$.

Beilinson shows that

$$r_{\mathcal{D}}\{\mathrm{div}(f),\mathrm{div}(g)\} = pr_{\mathcal{D}}(\log|f| \cup \log|g|) = [\mathrm{div}(\log|f|),\mathrm{div}(\log|g|)]_{\mathcal{D}}$$

and, according to example 3.2.8 , this equals

$$[\mathrm{div}(2\partial\log|f|),\mathrm{div}(2\partial\log|g|)]_{\mathcal{D}} = [\mathrm{div}(\frac{df}{f}),\mathrm{div}(\frac{dg}{g})]_{\mathcal{D}} =$$

$$= [\sum_i \mathrm{Res}_{x_i}(\frac{df}{f}).x_i, \sum_j \mathrm{Res}_{y_j}(\frac{dg}{g}).y_j] =$$

$$= [\mathrm{div}(f),\mathrm{div}(g)]_{\mathcal{D}} = \sum_{i,j} n_i m_j.[x_i,y_j]_{\mathcal{D}}.$$

Next, introducing the pairing $\langle\omega,\gamma\rangle = \frac{1}{2\pi i}\int_{E(\mathbf{C})}\gamma\wedge\omega$ between C^∞ 1-forms on $E(\mathbf{C})$, one finds after a short calculation:

$$\frac{1}{2\pi i}\int_{E(\mathbf{C})} r_{\mathcal{D}}(\{\mathrm{div}(f),\mathrm{div}(g)\})\wedge\omega = \langle\omega, pr_{\mathcal{D}}(\log|f|\cup\log|g|)\rangle =$$

$$= \frac{1}{2\pi i}\int_{E(\mathbf{C})}\log|f|\bar{\partial}\log|g|\wedge\omega\,.$$

For ω normalized such that $\frac{1}{2\pi i}\int_{E(\mathbf{C})}\bar{\omega}\wedge\omega = 1$ one has the analytic isomorphism

$$\phi: E(\mathbf{C}) \xrightarrow{\sim} \mathbf{C}/\Gamma\,,\quad \text{with } \phi(x) = \int_0^x \omega \bmod \Gamma\,,$$

which is $\mathrm{Gal}(\mathbf{C}/\mathbf{R})$-invariant.

Thus one has an identification of $(E(\mathbf{C}),\omega,F_\infty)$ with $(\mathbf{C}/\Gamma, dz, ^-)$. Here F_∞ denotes complex conjugation on $E(\mathbf{C})$ and $^-$ is complex conjugation on $\mathbf{C}$. The normalization condition for ω means that the volume of the lattice Γ is equal to π. Pontryagin duality pairs $\mathbf{C}/\Gamma$ and Γ as follows:

$$\begin{array}{rccc} (\ ,\ ): & \mathbf{C}/\Gamma\otimes\Gamma & \longrightarrow & \{z\in\mathbf{C}\mid |z|=1\}\subset\mathbf{C}^* \\ & (z,\gamma) & = & \exp(z\bar{\gamma}-\gamma\bar{z}) \end{array}$$

An explicit calculation can be carried out to give the result

$$\frac{1}{2\pi i}\int_{E(C}\log|f|\bar{\partial}\log|g|\wedge\omega = \langle\omega, pr_{\mathcal{D}}(\log|f|\cup\log|g|)\rangle =$$

$$= -\frac{1}{2}\sum_{i,j} n_i m_j \frac{d}{ds}E_1(x_i-y_j,s)\big|_{s=0} =$$

$$= -\frac{1}{2}\sum_{\substack{\gamma\in\Gamma\\ i,j}}{}' n_i m_j \frac{\bar{\gamma}}{|\gamma|^4}\overline{(x_i-y_j,\gamma)}\,. \tag{5.18}$$

Here we used $E_1(z,s)$ for the analytic continuation of the Eisenstein series

$$E_1(z,s) = \sum_{\gamma\in\Gamma}{}' \frac{(z+\gamma)}{|z+\gamma|^{2s}}\,,\quad \Re(s) > \frac{3}{2}\,.$$

More generally, one can define Eisenstein-Kronecker-Lerch (E-K-L) series $K_a(x,y,s)$, for $a \in \mathbf{Z}$ and $s \in \mathbf{C}$, by

$$K_a(x,y,s) = \sum_{\gamma\in\Gamma}{}' \frac{\chi_y(\gamma)\overline{(x+\gamma)}^a}{|x+\gamma|^{2s}},$$

with $\chi_y(\gamma) = \overline{(y,\gamma)}$. These E-K-L series admit a functional equation:

$$\Gamma(s)K_a(x,y,s) = \Gamma(1+a-s)K_a(y,x,a+1-s)\chi_y(-x).$$

The functional equation for $E/\mathbf{Q}$ then leads to relations of the form

$$L(E,s)_{s=2} \approx_{\mathbf{Q}^*} \sum_{i,j} n_i m_j K_1(0, x_i - y_j, 2). \tag{5.19}$$

A formula of this type comes as no surprise for elliptic curves $E/\mathbf{Q}$ with complex multiplication, because for such E, $L(E,s)$ is a Hecke L-function, and for this kind of L-functions the above result is known. However, the computer calculations of Bloch and Grayson were done for non-CM curves, but the results point to relations as those above. Of course, this agrees with Beilinson's conjecture, which was formulated for any elliptic curve.

Bloch and Grayson also suspected relations between the L-function $L(E,s)$ at the points $s = r = 3,4,5\ldots$ and the Eisenstein-Kronecker-Lerch series $K_1(0, x_i - y_j, r)$, $r \geq 3$, but results of C. Deninger seem to contradict this, at least in the non-CM case. However, taking the L-function of higher symmetric powers of $H^1(E)$, $\mathrm{Sym}^n H^1(E)$, $n \geq 1$, one should be able to find relations between a determinant of Eisenstein-Kronecker-Lerch series and $L(\mathrm{Sym}^n H^1(E), s)_{s=n+1}$.

We finish this chapter with a result of C. Deninger, (cf. [Den]).

Theorem 5.7.1 (Deninger) *Let $E/\mathbf{Q}$ be an elliptic curve with complex multiplication. Then, for every integer $r \geq 0$ the L-function $L(E,s)$ has a first order zero at $s = -r$ and one can construct an element*

$$\xi \in H^2_{\mathcal{M}}(E, \mathbf{Q}(2+r))_{\mathbf{Z}} = H^2_{\mathcal{M}}(E, \mathbf{Q}(2+r))$$

such that

$$r_{\mathcal{D}}(\xi) = L'(E,s)_{s=-r} \cdot \omega_{\mathbf{Q}} \in H^2_{\mathcal{D}}\left(E_{/\mathbf{R}}, \mathbf{R}(2+r)\right),$$

with a suitable generator $\omega_{\mathbf{Q}}$ *of the one-dimensional* $\mathbf{Q}$*-vector space* $H^1(E_{/\mathbf{R}}, \mathbf{Q}(1+r))$.

The proof depends on an explicit analytical formula for $L'(E,s)_{s=-r}$ and uses Deuring's result that $L(E,s) = L(\psi,s)$, where ψ is a Hecke character of E_K, K being the imaginary quadratic field of the complex multiplication.

Chapter 6

Beilinson's second conjecture

In this chapter the case of even i and $m = \frac{i}{2}$ is considered. It turns out that one must take into account the pole of the L-function at the point $s = m+1$. This leads to an old conjecture due to J. Tate and generalized by A. Beilinson. For Hilbert modular surfaces D. Ramakrishnan proved that part of motivic cohomology is enough to give a $\mathbf{Q}$–structure on Deligne cohomology with volume (up to a non-zero rational number) equal to the first non-zero coefficient of the Taylor series expansion of the L-function at $s = m$. This seems to be a general phenomenon.

6.1 Beilinson's Second Conjecture

Let i be even and $m = \frac{i}{2}$. In this case the regulator map

$$r_{\mathcal{D}} : H^{i+1}_{\mathcal{M}}(X, \mathbf{Q}(m+1))_{\mathbf{Z}} \longrightarrow H^{i+1}_{\mathcal{D}}(X_{/\mathbf{R}}, \mathbf{R}(m+1))$$

does not give a $\mathbf{Q}$–structure on Deligne cohomology.

Example 6.1.1 Let $X = \mathrm{Spec}(K)$, where K is a number field of degree $[K : \mathbf{Q}] = r_1 + 2r_2$. Dirichlet's Unit Theorem 1.3.1 says that

$$\dim_{\mathbf{Q}} H^1_{\mathcal{M}}(X, \mathbf{Q}(1))_{\mathbf{Z}} = \dim_{\mathbf{Q}} K_1(\mathcal{O}_K) \otimes \mathbf{Q} = r_1 + r_2 - 1$$

but on the other other hand,

$$\dim_{\mathbf{R}} H^1_{\mathcal{D}}(X_{/\mathbf{R}}, \mathbf{R}(1)) = r_1 + r_2 .$$

This value may be interpreted as

$$r_1 + r_2 = (r_1 + r_2 - 1) - (-1) = \operatorname{ord}_{s=0} \zeta_K(s) - \operatorname{ord}_{s=1} \zeta_K(s),$$

or, in other words, the difference of the dimensions is caused by the pole at $s = m = \frac{i}{2}$.

This idea was generalized to the case of arbitrary smooth, projective varieties over $\mathbf{Q}$ (or any number field) by Beilinson, thus relating it to a classical conjecture of Tate [Ta1] on the relation between the order of the L-function $L(X,s)$ at $s = \frac{i}{2}+1$ and the rank of the $\frac{i}{2}$-codimensional algebraic cycles on $X/\mathbf{Q}$ modulo homological equivalence.

Let $X/\mathbf{Q}$ be a smooth, projective variety of dimension d. Denote by $Z^m(X)$ the group of codimension m algebraic cycles on X. Then there are cycle class maps

$$cl_B : Z^m(X) \longrightarrow H^{2m}(X(\mathbf{C}), \mathbf{Q}(m)),$$

where $cl_B(Z)$ is $(2\pi i)^m$ times the Poincaré dual of the topological $(2d-2m)$-cycle defined by Z. One also has the ℓ-adic version

$$cl_\ell : Z^m(X) \longrightarrow H^{2m}_{\acute{e}t}(X \otimes \bar{\mathbf{Q}}, \mathbf{Q}_\ell(m)),$$

which is invariant under the action of $\mathrm{Gal}(\bar{\mathbf{Q}}/\mathbf{Q})$. These cycle class maps are related to each other by the comparison isomorphism

$$\alpha_m : H^{2m}(X(\mathbf{C}), \mathbf{Q}(m)) \otimes \mathbf{Q}_\ell \xrightarrow{\sim} H^{2m}_{\acute{e}t}(X \otimes \bar{\mathbf{Q}}, \mathbf{Q}_\ell(m)).$$

The **Chow group** of codimension m cycles is defined as $Z^m(X)$ modulo rational equivalence. Here rational equivalence for $m \geq 1$ is a generalization of linear equivalence in the case of divisors. More precisely, two cycles $V, W \in Z^m(X)$ are called **rationally equivalent** if there are closed irreducible codimension $m-1$ subvarieties $Y_1, \ldots, Y_s$ and functions $f_1, \ldots, f_s$ such that $V - W = \sum_i \operatorname{div}(f_i)$, where for $Z \subset Y_i$, a closed, irreducible divisor in Y_i,

$$\operatorname{div}(f_i) = \sum_{Z \subset Y_i} \operatorname{length}(\mathcal{O}_{Y_i,Z}/f_i\mathcal{O}_{Y_i,Z}).Z\,.$$

Thus the mth Chow group $CH^m(X) = Z^m(X)/\text{rat.eq.}$. The cycle class maps cl_B and cl_ℓ factor over $CH^m(X)$. A cycle class $Z \in CH^m(X) \otimes \mathbf{Q}$ is called **homologically trivial** if $cl_B(Z) = cl_\ell(Z) = 0$. This is well defined by the injectivity, for any ℓ, of the maps

$$H^{2m}(X(\mathbf{C}), \mathbf{Q}(m)) \longrightarrow H^{2m}_{\acute{e}t}(X \otimes \bar{\mathbf{Q}}, \mathbf{Q}_\ell(m))\,.$$

Two cycles $V, W \in Z^m(X)$ are **homologically equivalent** if their difference is homologically trivial. Homological equivalence defines another equivalence relation on $Z^m(X)$, coarser than rational equivalence. We write $B^m(X) = Z^m(X)/hom.eq.$, and also $A^m(X)$ for the homologically trivial cycles in $Z^m(X)$. We use the same notation for cycle classes modulo rational equivalence, in other words, we also write

$$A^m(X) = \operatorname{Ker}(CH^m(X) \xrightarrow{cl_B} H^{2m}(X(\mathbf{C}), \mathbf{Q}(m)))$$

and

$$B^m(X) = \operatorname{Im}(CH^m(X) \xrightarrow{cl_B} H^{2m}(X(\mathbf{C}), \mathbf{Q}(m)))\,.$$

For the ℓ-adic version one has

$$A^m(X) = \operatorname{Ker}(CH^m(X) \xrightarrow{cl_\ell} H^{2m}_{\acute{e}t}(X \otimes \bar{\mathbf{Q}}, \mathbf{Q}_\ell(m))^G)$$

and

$$B^m(X) = \operatorname{Im}(CH^m(X) \xrightarrow{cl_\ell} H^{2m}_{\acute{e}t}(X \otimes \bar{\mathbf{Q}}, \mathbf{Q}_\ell(m))^G)\,,$$

where G=Gal($\bar{\mathbf{Q}}/\mathbf{Q}$). Thus one gets a short exact sequence

$$0 \longrightarrow A^m(X) \longrightarrow CH^m(X) \longrightarrow B^m(X) \longrightarrow 0\,. \tag{6.1}$$

For $X/\mathbf{Q}$ smooth, projective let $L(X,s)$ denote the L-function defined by the cohomology H^i with $i = 2m$. Then one has the following conjecture, (cf. [Ta1]).

Conjecture 6.1.1 (Tate) $\operatorname{ord}_{s=m+1} L(X,s) = -\operatorname{rank}(B^m(X))\,.$

It is known that $cl_{\mathcal{B}}(B^m(X)) \subset H^{2m}(X(\mathbf{C}), \mathbf{Q}(m)) \cap H^{m,m}(X(\mathbf{C}))$ and one deduces an injection

$$z_{\mathcal{D}} : B^m(X) \longrightarrow H_{\mathcal{D}}^{2m+1}(X_{/\mathbf{R}}, \mathbf{R}(m+1)).$$

We can state Beilinson's second conjecture:

Conjecture 6.1.2 (Beilinson II) *For* $m = \frac{i}{2}$,
(*i*) $r_{\mathcal{D}} \oplus z_{\mathcal{D}} : H_{\mathcal{M}}^{i+1}(X, \mathbf{Q}(m+1))_{\mathbf{Z}} \oplus (B^m(X) \otimes \mathbf{Q}) \longrightarrow$
$\longrightarrow H_{\mathcal{D}}^{i+1}(X_{/\mathbf{R}}, \mathbf{R}(m+1))$
induces a $\mathbf{Q}$*–structure on Deligne cohomology.*
(*ii*) $\operatorname{ord}_{s=m} L(X,s) = \dim_{\mathbf{Q}} H_{\mathcal{M}}^{i+1}(X, \mathbf{Q}(m+1))_{\mathbf{Z}}$.
(*iii*) $\operatorname{ord}_{s=m+1} L(X,s) = -\operatorname{rank}(B^m(X))$ (Tate).
(*iv*) *The volume* $\det(r_{\mathcal{D}} \oplus z_{\mathcal{D}})$ *defined by the* $\mathbf{Q}$*–structure in* (*i*) *satisfies:*

$$\det(r_{\mathcal{D}} \oplus z_{\mathcal{D}}) \approx_{\mathbf{Q}^*} L^*(X,s)_{s=m}.$$

Example 6.1.2 $X = \operatorname{Spec}(K)$, K a number field, and $i = 0$.

6.2 Hilbert Modular Surfaces

An interesting situation arises when X is a Hilbert modular surface over $\mathbf{Q}$ (and actually over any abelian extension of $\mathbf{Q}$), $i = 2$ and $m = 1$. First, for any smooth, projective variety X over $\mathbf{Q}$, the short exact sequence

$$0 \longrightarrow A^1(X) \longrightarrow CH^1(X) \longrightarrow B^1(X) \longrightarrow 0$$

can be written as

$$0 \longrightarrow \operatorname{Pic}^0(X) \longrightarrow \operatorname{Pic}(X) \longrightarrow NS(X) \longrightarrow 0,$$

where $\operatorname{Pic}(X) = \operatorname{Pic}(X \otimes \bar{\mathbf{Q}})^G$ denotes the group of $\mathbf{Q}$–rational divisors on X modulo linear equivalence, and $NS(X)$ is the Néron-Severi group of X, i.e. the group of $\mathbf{Q}$–rational divisors on X modulo algebraic equivalence. In general algebraic equivalence lies between rational and homological equivalence, but for the 1-codimensional (divisorial) case it coincides with homological equivalence. The Néron-Severi group is finitely generated. The group $\mathrm{Pic}^0(X)$ has the structure of an abelian variety of dimension equal to $\dim H^1(X, \mathcal{O}_X)$.

Now let F be a real quadratic number field and let G denote the Weil restriction $R_{F/\mathbf{Q}}\mathrm{GL}_2/F$. Thus G is a $\mathbf{Q}$-group and one has e.g. $G(\mathbf{Q}) = \mathrm{GL}_2(F)$ and $G(\mathbf{R}) = \mathrm{GL}_2(\mathbf{R}) \times \mathrm{GL}_2(\mathbf{R})$. Also, let $\mathbf{A} = \mathbf{R} \times \mathbf{A}_f$, $\mathbf{A}_f = \prod_p \mathbf{Z}_p \otimes \mathbf{Q}$, denote the $\mathbf{Q}$-adèles. Then one can construct a quasi-compact, separated scheme S over $\mathbf{Q}$ with a continuous $G(\mathbf{A}_f)$-action, such that the set of its complex points $S(\mathbf{C})$ is equal to $G(\mathbf{Q})\backslash G(\mathbf{A})/K_\infty$, where $K_\infty = L_\infty \times L_\infty$, and

$$L_\infty = \left\{ \begin{pmatrix} a & -b \\ b & a \end{pmatrix} \in \mathrm{GL}_2(\mathbf{R}) \right\}.$$

One may compactify S to obtain $\bar{S} = S \cup S^\infty$. Furthermore, for each compact, open subgroup K of $G(\mathbf{A}_f)$ one has a surface $S_K = S/K$ defined over $\mathbf{Q}$, such that $S = \varprojlim S_K$, the limit being taken over K. Adding a finite set of 'cusps' S_K^∞, one has the so-called Baily-Borel-Satake compactification $\bar{S}_K = S_K \cup S_K^\infty$, which can be embedded as a closed, normal subvariety of a suitable projective space. This $\bar{S}_K$ admits a smooth resolution over $\mathbf{Q}$,

$$\sigma : \tilde{S}_K \longrightarrow \bar{S}_K .$$

This desingularized $\tilde{S}_K$ is called a Hilbert modular surface. By work of M. Rapoport such Hilbert modular surfaces admit models over $\mathbf{Z}$.

The S_k's arise as moduli spaces of polarized abelian surfaces with some additional properties. We refer to [vdG] for details on Hilbert modular surfaces.

The following nice result is due to G. Harder, R. Langlands and M. Rapoport:

Theorem 6.2.1 (Harder, Langlands, Rapoport) *Tate's Conjecture is true for Hilbert modular surfaces $\tilde{S}_K$ over any abelian extension of $\mathbf{Q}$, and any compact open subgroup $K \subset G(\mathbf{A}_f)$.*

Hilbert modular surfaces admit many non-trivial divisors, the so-called Hirzebruch-Zagier cycles, which emerge as translates of a diagonal modular curve on the surface. To understand how these cycles arise, let $H = \mathrm{GL}_2/\mathbf{Q}$ and denote by $\Delta : H \longrightarrow G$ the natural diagonal map of $\mathbf{Q}$-groups. H defines a modular curve M over $\mathbf{Q}$, whose set of $\mathbf{C}$-points is given by

$$M(\mathbf{C}) = H(\mathbf{Q})\backslash H(\mathbf{A})/L_\infty .$$

One can compactify M to obtain $\bar{M} = M \cup M^\infty$ such that $\bar{M} = \varprojlim M_L$, where $M_L = M/L$ for any compact open subgroup $L \subset H(\mathbf{A}_f)$. Δ induces embeddings $\Delta^* : M \longrightarrow S$, $M^\infty \longrightarrow S^\infty$ and $\bar{M} \longrightarrow \bar{S}$. $\bar{S}$ admits a right action ρ of $G(\mathbf{A}_f)$. Then, writing $\pi_K : \bar{S} \longrightarrow \bar{S}_K$ for the natural projection over $\mathbf{Q}$ such that $\pi_K(S) = S_K$ and $\pi_K(S^\infty) = S_K^\infty$ for any compact open $K \subset G(\mathbf{A}_f)$, one defines, for $g \in G(\mathbf{A}_f)$:

$$\begin{cases} \mathcal{C}(g,K) &= \pi_K(\rho(g)(\Delta^*(M))) \subset S_K , \quad \text{and} \\ \bar{\mathcal{C}}(g,K) &= \pi_K(\rho(g)(\Delta^*(\bar{M}))) \subset \bar{S}_K . \end{cases}$$

A Hirzebruch-Zagier cycle $\tilde{\mathcal{C}}(g,K)$ on $\tilde{S}_K$ is now defined as the closure of $\mathcal{C}(g,K)$ in $\tilde{S}_K$. One also uses this terminology for the $\mathcal{C}(g,K)$ on S_K and for the $\bar{\mathcal{C}}(g,K)$ on $\bar{S}_K$. There is a non-trivial morphism over $\mathbf{Q}$

$$\lambda(g,K) : \bar{M}_L \longrightarrow \bar{\mathcal{C}}(g,K) \quad \text{with} \quad L = H(\mathbf{A}_f) \cap gKg^{-1} .$$

A modular unit on $\bar{\mathcal{C}}(g,K) \otimes \bar{\mathbf{Q}}$ is a function f with $f \circ \lambda(g,K) \in \mathcal{O}^*_{\bar{M}_L \otimes \bar{\mathbf{Q}}}$, so the zeroes and poles of $f \circ \lambda$ lie at the cusps of $\bar{M}_L$. A finite formal sum $\sum_j (\mathcal{C}_j, f_j)$ is called K-admissible if

(i) each $\mathcal{C}_j$ is a cycle of the form $\mathcal{C}(g,K) \otimes \bar{\mathbf{Q}}$;

(ii) each f_j is a modular unit on $\mathcal{C}_j$;

(iii) $\sum_j \mathrm{div}(f_j) = 0$ as a zero cycle on $\bar{S}_K \otimes \bar{\mathbf{Q}}$.

Define $R'_K \subset H^3_{\mathcal{M}}(\tilde{S}_K, \mathbf{Q}(2))$ as the $\mathbf{Q}$-subspace generated by the K-admissible sums $\sum_j (\mathcal{C}_j, f_j)$. For $N \subset K$ an open subgroup, one has a natural morphism

$$\tau_{K/N} : H^3_{\mathcal{M}}(\tilde{S}_N, \mathbf{Q}(2)) \longrightarrow H^3_{\mathcal{M}}(\tilde{S}_K, \mathbf{Q}(2))$$

and one finally defines the $\mathbf{Q}$-subspace $R_K \subset H^3_{\mathcal{M}}(\tilde{S}_K, \mathbf{Q}(2))$ as the subspace generated by the $\{\tau_{K/N}(R'_K) | N \subset K \text{ open}\}$.

Concerning the integral nature of R_K one may observe that there exists a proper model $\mathcal{C}_{\mathbf{Z}}$ over Spec($\mathbf{Z}$) of a finite union of Hirzebruch-Zagier cycles $\mathcal{C}$ because the latter are translates of modular curves, which admit a regular, proper model over Spec($\mathbf{Z}$). Also, modular units extend to the model over Spec($\mathbf{Z}$).

Let $\mathcal{C}$ be a finite union of Hirzebruch-Zagier cycles on $\tilde{S}_K$ and let $\mathcal{C}_{\mathbf{Z}}$ be a proper model of $\mathcal{C}$ over Spec($\mathbf{Z}$). Denote by

$$\imath : \mathcal{C}_{\mathbf{Z}} \longrightarrow \mathcal{C} \longrightarrow \tilde{S}_K$$

the natural map. D. Ramakrishnan proved, (cf. [Ra2]):

$$R_K \subset \operatorname{Im}(K_1'(\mathcal{C}_{\mathbf{Z}}) \xrightarrow{\imath_*} K_1(\tilde{S}_K)) \otimes \mathbf{Q}\,.$$

The main result with respect to Beilinson's second conjecture, due to Ramakrishnan, is the following theorem.

Theorem 6.2.2 (Ramakrishnan) *Let $K \subset G(\mathbf{A}_f)$ be a compact open subgroup and let $\tilde{S}_K$ be the corresponding Hilbert modular surface. Then:*

(i) $r_{\mathcal{D}} \oplus z_{\mathcal{D}} : R_K \oplus (NS(\tilde{S}_K) \otimes \mathbf{Q}) \longrightarrow H^3_{\mathcal{D}}(\tilde{S}_{K/\mathbf{R}}, \mathbf{R}(2))$ *defines a $\mathbf{Q}$-structure on Deligne cohomology.*

(ii) $\det(r_{\mathcal{D}} \oplus z_{\mathcal{D}}) \approx_{\mathbf{Q}^*} L^*(\tilde{S}_K, s)_{s=1}$.

Remark 6.2.1 Actually the theorem was proved in greater generality, just as Tate's conjecture, for abelian number fields over $\mathbf{Q}$.

Chapter 7

Arithmetic intersections and Beilinson's third conjecture

This chapter treats the case of odd i and $m = \frac{i+1}{2}$. The L-functions considered in the sequel will be those defined by the cohomology space H^i. Beilinson's third conjecture regards this situation for smooth, projective varieties over $\mathbf{Q}$, and reduces to a weakened form of the Birch & Swinnerton-Dyer Conjectures in the case of an elliptic curve or an abelian variety over $\mathbf{Q}$. The elliptic regulator is generalized to become the determinant of an arithmetic intersection index on arithmetic varieties on Spec($\mathbf{Z}$), *thus enlarging Arakelov's construction of the Néron-Tate height pairing. This generalized height pairing was constructed by Beilinson and, independently, by Gillet and Soulé.*

7.1 The Intersection Pairing

The construction goes as follows. Let X be a smooth, projective variety over $\mathbf{Q}$ of dimension $d-1$, and assume that X admits a regular arithmetic model $X_{\mathbf{Z}}$ over Spec($\mathbf{Z}$) with so-called Arakelov compactification $\bar{X}_{\mathbf{Z}} = (X_{\mathbf{Z}}, \omega)$, where ω is a real Kähler $(1,1)$-form on $X(\mathbf{C})$ such that $F_\infty^* \omega = -\omega$. Here F_∞ is the orientation reversing involution on $X(\mathbf{C})$ induced by complex conjugation. For an integral codimension $p-1$ subscheme $W \overset{i}{\hookrightarrow} X_{\mathbf{Z}}$ and a non-zero rational function f on W, one may consider $\log|f|$ as an L^1-function on the non-singular locus of

$W(\mathbf{C}) \hookrightarrow X(\mathbf{C})$ (if $W(\mathbf{C})$ is not empty), giving rise to a current $\imath_* \log|f|$ on $X(\mathbf{C})$. Let $H(\imath_* \log|f|)$ be its harmonic $(p-1,p-1)$-projection (zero in case $W(\mathbf{C})$ is empty). Then $H(\imath_* \log|f|) \in \mathcal{H}^{p-1,p-1}(X_{/\mathbf{R}})$.

Definition 7.1.1 *The* Arakelov divisor $\overline{\mathrm{div}}(f)$ *of* f *is the pair*

$$(\mathrm{div}(f), -H(\imath_* \log|f|)) \in Z^p(\bar{X}_{\mathbf{Z}}),$$

where $Z^p(\bar{X}_{\mathbf{Z}})$ *consists of pairs* (ξ, α) *of codimension* p *cycles* ξ *on* $X_{\mathbf{Z}}$ *and real harmonic forms* α *of type* $(p-1,p-1)$ *on* $X(\mathbf{C})$, *with* $F_\infty^* \alpha = (-1)^{p-1}\alpha$.

Definition 7.1.2 *The* arithmetic Chow group $CH^p(\bar{X}_{\mathbf{Z}})$ *is the quotient of* $Z^p(\bar{X}_{\mathbf{Z}})$ *by the subgroup of all cycles of the form* $\overline{\mathrm{div}}(f)$.

Example 7.1.1 Let $\mathcal{L}$ be a line bundle on X with induced holomorphic line bundle $\mathcal{L}(\mathbf{C})$ on $X(\mathbf{C})$. Also let $<\ ,\ >$ be a hermitian metric on $\mathcal{L}(\mathbf{C})$ with associated norm $|.| = \sqrt{<\ ,\ >}$. Then $|.|$ is called admissible (for $\bar{X}_{\mathbf{Z}}$) if it is F_∞-invariant and has harmonic curvature form. Every such $\mathcal{L}$ admits an admissible metric. A **metrized line bundle** $\bar{\mathcal{L}}$ on $\bar{X}_{\mathbf{Z}}$ consists of a pair $(\mathcal{L}, |.|)$ with $|.|$ admissible on $\mathcal{L}(\mathbf{C})$. An isomorphism $\bar{\mathcal{L}} \xrightarrow{\sim} \bar{\mathcal{L}}'$ is an isomorphism $\mathcal{L} \xrightarrow{\sim} \mathcal{L}'$ which respects the norms. $\mathrm{Pic}(\bar{X}_{\mathbf{Z}})$ denotes the group of isomorphism classes of metrized line bundles on $X_{\mathbf{Z}}$, the product being given by tensor product.

With the notations of the above example, the following holds:

Theorem 7.1.1 $CH^1(\bar{X}_{\mathbf{Z}})$ *and* $\mathrm{Pic}(\bar{X}_{\mathbf{Z}})$ *are canonically isomorphic.*

Using Quillen's spectral sequence (cf. Theorem 4.4.5) one gets a complex

$$\coprod_{x \in X_{\mathbf{Z}}^{(p-2)}} K_2(k(x)) \xrightarrow{d_{p-1}} \coprod_{y \in X_{\mathbf{Z}}^{(p-1)}} k(y)^* \xrightarrow{d_p} Z^p(X_{\mathbf{Z}}) = \coprod_{z \in X_{\mathbf{Z}}^{(p)}} \mathbf{Z} \longrightarrow 0,$$

where $K_2(k(x))$ is Milnor's K_2 and d_{p-1} is essentially a sum of tame symbols. Define $CH^{p,p-1}(X_{\mathbf{Z}}) = \mathrm{Ker}(d_p)/\mathrm{Im}(d_{p-1})$, then there is an exact sequence for $p \geq 0$:

$$CH^{p,p-1}(X_{\mathbf{Z}}) \xrightarrow{r} \mathcal{H}^{p-1,p-1}(X_{/\mathbf{R}}) \xrightarrow{i} CH^p(\bar{X}_{\mathbf{Z}}) \xrightarrow{\epsilon} CH^p(X_{\mathbf{Z}}) \longrightarrow 0. \tag{7.1}$$

Here r is the 'regulator map' (e.g. take $X = \mathrm{Spec}(K)$, K a number field, and $p = 1$) and $i(\alpha) = (0, \alpha)$ and $\varepsilon(\xi, \alpha) = \xi$.

Gillet and Soulé constructed a bi-additive and symmetric intersection pairing

$$\langle\, , \,\rangle : CH^p(\bar{X}_{\mathbf{Z}}) \otimes CH^{d-p}(\bar{X}_{\mathbf{Z}}) \longrightarrow \mathbf{R}$$

as the sum of four terms

$$\langle(\xi, \alpha), (\eta, \beta)\rangle = \langle\xi, \eta\rangle_f + \langle\xi, \eta\rangle_\infty + \langle\xi, \beta\rangle + \langle\alpha, \eta\rangle\,,$$

thus $\langle\alpha, \beta\rangle = 0$. On the level of cycles their construction goes as explained in the following lines.

Let $\bar{Z}_1$ be represented by

$$(Z_1, \alpha) \in Z^p(\bar{X}_{\mathbf{Z}}) = Z^p(X_{\mathbf{Z}}) \oplus \mathcal{H}^{p-1,p-1}(X_{/\mathbf{R}})\,,$$

and let $\bar{Z}_2$ be represented by

$$(Z_2, \beta) \in Z^{d-p}(\bar{X}_{\mathbf{Z}}) = Z^{d-p}(X_{\mathbf{Z}}) \oplus \mathcal{H}^{d-p-1,d-p-1}(X_{/\mathbf{R}})$$

and assume that the supports of Z_1 and Z_2 have empty intersection on $X/\mathbf{Q}$, the generic fibre. Then one defines the **intersection number**

$$\begin{aligned}\langle(Z_1, \alpha), (Z_2, \beta)\rangle &= \sum_p \langle Z_1, Z_2\rangle_p + \langle Z_1, Z_2\rangle_\infty + \\ &\quad + \langle Z_1, \beta\rangle + \langle\alpha, Z_2\rangle\,. \qquad (7.2)\end{aligned}$$

Let's explain the terms occuring in this sum.

First, one has for the classes of the structure sheaves of Z_1 and Z_2, in the K-theory of $X_{\mathbf{Z}}$ with supports, the product

$$[\mathcal{O}_{Z_1}].[\mathcal{O}_{Z_2}] \in K_0^{\mathrm{Supp}(Z_1)\cap\mathrm{Supp}(Z_2)}(X_{\mathbf{Z}})$$

with $\mathrm{Supp}(Z_1) \cap \mathrm{Supp}(Z_2)$ contained in a finite union of closed fibres X_p over the primes p. Write π for the structure map

$$\pi : X_{\mathbf{Z}} \longrightarrow \mathrm{Spec}(\mathbf{Z})\,,$$

then $\mathrm{ord}_p\langle Z_1, Z_2\rangle$ is the image of $[\mathcal{O}_{Z_1}].[\mathcal{O}_{Z_2}]$ in $\mathbf{Z}$ under the composed morphism

$$K_0^{\mathrm{Supp}(Z_1)\cap\mathrm{Supp}(Z_2)}(X_{\mathbf{Z}}) \longrightarrow K_0^{X_p}(X_{\mathbf{Z}}) = K_0'(X_p) \xrightarrow{\pi_*} K_0(\mathbf{F}_p) = \mathbf{Z}\,,$$

and $\langle Z_1, Z_2\rangle_p = -\mathrm{ord}_p\langle Z_1, Z_2\rangle \log(p)$.

Next, let δ_Z, Z a cycle on $X_{\mathbf{Z}}$, denote the current of integration over $Z(\mathbf{C})$, then $\langle Z_1, \beta\rangle = \delta_{Z_1}(\beta)$ and $\langle \alpha, Z_2\rangle = \delta_{Z_2}(\alpha)$. Finally

$$\langle Z_1, Z_2\rangle_\infty = -\int_{Z_2(\mathbf{C})} g_{Z_1}\,,$$

where g_{Z_1} is the current on $X_{/\mathbf{R}}$ of type $(p-1, p-1)$ which is a solution of the differential equation

$$\frac{1}{\pi i}\partial\bar\partial g_{Z_1} = \delta_{Z_1} - H(\delta_{Z_1})\,,$$

and such that $H(g_{Z_1}) = 0$. One may show that this pairing respects rational equivalence and therefore the following theorem, due to Gillet and Soulé, can be formulated:

Theorem 7.1.2 (Gillet, Soulé) *The intersection number*

$$\langle (Z_1, \alpha), (Z_2, \beta)\rangle$$

defines a bi-additive and symmetric pairing

$$CH^p(\bar X_{\mathbf{Z}}) \otimes CH^{d-p}(\bar X_{\mathbf{Z}}) \longrightarrow \mathbf{R}\,.$$

Actually, for $d = 2$ and $p = 1$ one recovers Arakelov's intersection index on an arithmetic surface.

Write

$$A^p(X_{\mathbf{Z}}) = \mathrm{Ker}(CH^p(\bar X_{\mathbf{Z}}) \longrightarrow H^{2p}(X(\mathbf{C}), \mathbf{Q}(p)))/\mathcal{H}^{p-1,p-1}(X_{/\mathbf{R}})$$

for the subgroup of $CH^p(X_{\mathbf{Z}})$ consisting of cycles homologically trivial on $X(\mathbf{C})$, then one obtains a pairing, independent of the choice of the Kähler metric on $X(\mathbf{C})$:

$$A^p(X_{\mathbf{Z}}) \otimes A^{d-p}(X_{\mathbf{Z}}) \longrightarrow \mathbf{R}\,. \tag{7.3}$$

Beilinson uses a somewhat different notation. He defines the group

$$\begin{aligned} CH^p(X)^0 &= \operatorname{Ker}(CH^p(X) \longrightarrow H^{2p}_{\acute{e}t}(X \otimes \bar{\mathbf{Q}}, \mathbf{Q}_\ell(p))) \\ &= \operatorname{Ker}(CH^p(X) \longrightarrow H^{2p}(X(\mathbf{C}), \mathbf{Q}(p))), \end{aligned} \tag{7.4}$$

and conjectures this to be equivalent to

$$CH^p(X)^0 = \operatorname{Im}(CH^p(X_{\mathbf{Z}})^0 \longrightarrow CH^p(X)), \tag{7.5}$$

where $CH^p(X_{\mathbf{Z}})^0 \subset CH^p(X_{\mathbf{Z}})$ is the subgroup of cycles (modulo rational equivalence) whose intersection with the fibres of $X_{\mathbf{Z}}$ is homologous to zero for any prime. One expects that

$$CH^p(X)^0 \otimes \mathbf{Q} \xrightarrow{\sim} A^p(X_{\mathbf{Z}}) \otimes \mathbf{Q}. \tag{7.6}$$

For two cycles Z_1 and Z_2 on X of codimension p and $d-p$, respectively, with classes $cl(Z_1) \in CH^p(X)^0$ and $cl(Z_2) \in CH^{d-p}(X)^0$, and such that $\operatorname{Supp}(Z_1) \cap \operatorname{Supp}(Z_2) = \emptyset$, Beilinson defines the **intersection number**

$$\langle Z_1, Z_2 \rangle \in \mathbf{R}$$

as the sum of local intersection numbers. This constuction is equivalent to the one of Gillet and Soulé. Beilinson's result can be stated as follows, (cf. [Be4]):

Theorem 7.1.3 (Beilinson) *There exists a unique bilinear pairing*

$$\langle\ ,\ \rangle : CH^p(X)^0 \otimes CH^{d-p}(X)^0 \longrightarrow \mathbf{R},$$

such that $\langle cl(Z_1), cl(Z_2)\rangle = \langle Z_1, Z_2\rangle$ *if the supports of* Z_1 *and* Z_2 *do not intersect.*

Tensoring with $\mathbf{Q}$, one may write

$$H^{2p}_{\mathcal{M}}(X, \mathbf{Q}(p))^0_{\mathbf{Z}} = CH^p(X)^0 \otimes \mathbf{Q}$$

and one obtains a pairing

$$\langle\ ,\ \rangle : H^{2p}_{\mathcal{M}}(X, \mathbf{Q}(p))^0_{\mathbf{Z}} \otimes H^{2(d-p)}_{\mathcal{M}}(X, \mathbf{Q}(d-p))^0_{\mathbf{Z}} \longrightarrow \mathbf{R}. \tag{7.7}$$

7.2 Beilinson's Third Conjecture

Between the lines and in agreement with a variant of a generalized Birch & Swinnerton-Dyer Conjecture it is believed that the $H^{2k}_{\mathcal{M}}(X, \mathbf{Q}(k))^0_{\mathbf{Z}}$ are finite dimensional spaces, so it should make sense to define the determinant $\det\langle\ ,\ \rangle$ with respect to some $\mathbf{Q}$–bases of both motivic cohomology spaces involved in the intersection pairing as described above. Moreover, one has an isomorphism

$$F^m H^{2m-1}_{DR}(X_{/\mathbf{R}}) \xrightarrow{\sim} H^{2m-1}(X_{/\mathbf{R}}, \mathbf{R}(m-1)),$$

giving a period matrix Π. Combining the above constructions and conjectures, one can state Beilinson's third conjecture:

Conjecture 7.2.1 (Beilinson III) *Let X be a smooth, projective variety defined over $\mathbf{Q}$, and assume that X has a regular, proper model $X_{\mathbf{Z}}$ over* $\mathrm{Spec}(\mathbf{Z})$. *Then, for $m = \frac{i+1}{2}$:*

(i) $\dim_{\mathbf{Q}} H^{2m}_{\mathcal{M}}(X, \mathbf{Q}(m))^0_{\mathbf{Z}} < \infty$;

(ii) $\langle\ ,\ \rangle$ *is a non-degenerate pairing* ;

(iii) $\mathrm{ord}_{s=m} L(X,s) = \dim_{\mathbf{Q}} H^{2m}_{\mathcal{M}}(X, \mathbf{Q}(m))^0_{\mathbf{Z}}$;

(iv) $L(X,s)_{s=m} \approx_{\mathbf{Q}^*} \det \Pi \det\langle\ ,\ \rangle^{\cdot}$.

Example 7.2.1 For an abelian variety $X/\mathbf{Q}$ and $i = m = 1$ part *(i)* of the conjecture gives the Mordell-Weil Theorem.

Example 7.2.2 Let $X/\mathbf{Q}$ be an elliptic curve with (regular) Néron model $\mathcal{E} = X_{\mathbf{Z}}$. Then $H^2_{\mathcal{M}}(X, \mathbf{Q}(1)) = H^2_{\mathcal{M}}(\mathcal{E}, \mathbf{Q}(1))$ and the rank of $X(\mathbf{Q})$,

$$\mathrm{rank}(X(\mathbf{Q})) = \dim_{\mathbf{Q}} H^2_{\mathcal{M}}(X, \mathbf{Q}(1))^0 = \dim_{\mathbf{Q}} H^2_{\mathcal{M}}(X, \mathbf{Q}(1))^0_{\mathbf{Z}}.$$

Part *(iii)* corresponds to the first part of the Birch & Swinnerton-Dyer Conjectures. Parts *(ii)* and *(iv)* are a weak version of the second part of the Birch & Swinnerton-Dyer Conjectures for X. Also, cf. the Appendix to Chapter 2.

Example 7.2.3 Part *(iii)* for arbitrary smooth, projective varieties over $\mathbf{Q}$ (or any number field) is just Bloch's 'Recurring Fantasy', (cf. [Bl2]).

Remark 7.2.1 In general one has no idea of how possibly missing rational factors in Beilinson's Conjectures should look like, and in fact, there is no hope to find them. This follows from the fact that the definition of the motivic cohomology in the conjectures, in terms of the γ-filtration on the K-groups, is well defined only after tensoring with $\mathbf{Q}$. Bloch's generalized Chow groups $CH^\bullet(X,\star)$, cf. the end of Chapter 3, which are integrally defined may help to sharpen the regulator map to an Abel–Jacobi map from $CH^\bullet(X,\star)$ to a suitable intermediate Jacobian. A relation between the regulator map and the Abel-Jacobi map also occurs in U. Jannsen's work to be discussed in the next chapter.

To close this chapter and the previous ones on the formulation of Beilinson's conjectures we give the picture showing the values of the argument s of the L-function $L(X,s)$ defined by the various cohomology spaces H^i and the corresponding Beilinson Conjecture (if these values make sense as integers):

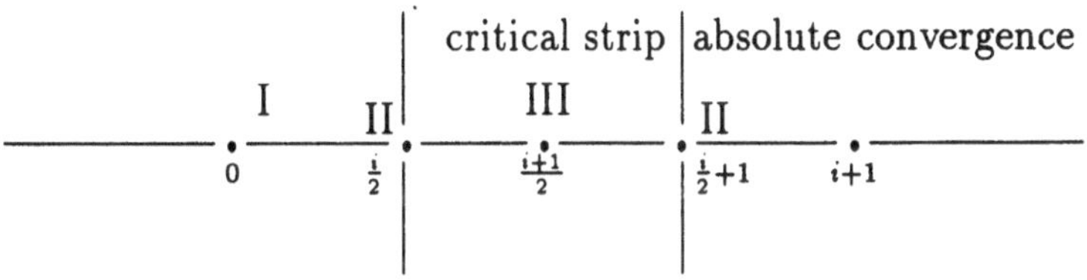

Chapter 8

Absolute Hodge cohomology, Hodge and Tate conjectures and Abel-Jacobi maps

Absolute Hodge cohomology is presented as a Poincaré duality theory that generalizes Deligne-Beilinson cohomology in the sense that it includes the weight filtration. In this way it applies to general schemes over the complex numbers. The relation with motivic cohomology is again given by a regulator map that is conjectured to have dense image, at least for smooth schemes that can be defined over a number field. This conjectured property induces the classical Hodge Conjecture for smooth, projective varieties.

The ℓ-adic counterpart of the Hodge conjecture is Tate's conjecture on algebraic cycles. The parallelism becomes transparent when one uses the language of motives. Also, for smooth varieties over number fields, cohomologically trivial cycles suggest the existence of 'motivic Abel-Jacobi maps' into an extension group of (mixed) motives. In a special situation the injectivity of these Abel-Jacobi maps implies the surjectivity of corresponding regulator maps. These ideas are due to U. Jannsen.

8.1 The Hodge Conjecture

Let X be a smooth, projective variety over a field k, and let $H^i(X)$ be a suitable cohomology space, e.g. when $k = \mathbf{Q}$, $H^i(X)$ may be taken to be the motivic cohomology space $H^i_{\mathcal{M}}(X, \mathbf{Q}(j))$, or when $k = \mathbf{R}$ or $\mathbf{C}$, $H^i(X)$ may be Deligne cohomology $H^i_{\mathcal{D}}(X, \mathbf{R}(j))$. Then one defines the **filtration by coniveau**

$N^pH^i(X) =$

$$= \bigcup_Z \{\mathrm{Ker}(H^i(X) \longrightarrow H^i(X\backslash Z)) | Z \subset X \text{ closed of codimension } \geq p\}$$

$$= \bigcup_Z \{\mathrm{Im}(H^i_Z(X) \longrightarrow H^i(X)) | Z \subset X \text{ closed of codimension } \geq p\}. \tag{8.1}$$

$H^i(X)$ is said to have **support in codimension** $\geq p$ if $N^pH^i(X) = H^i(X)$, in other words $H^i(X)$ has support in codimension p if every $x \in H^i(X)$ lies in the image of $H^i_Z(X)$ for some closed subvariety $Z \subset X$ of codimension p. Of course this must make sense in the underlying cohomology theory.

The filtration by coniveau defines a spectral sequence whose E_1-term was introduced by A. Grothendieck (cf. [Gr1]) in 1963 as

$$E_1^{p,q} = \coprod_{x \in X^{(p)}} H^{q-p}(k(x)),$$

where $H^\bullet(k(x))$ is the de Rham cohomology of $k(x)/k$. For k of characteristic zero and $H^i(X)$ the de Rham cohomology, the spectral sequence abuts to $H^\bullet(X)$. The differentials

$$d_1^{p,q} : E_1^{p,q} \longrightarrow E_1^{p+1,q}$$

give rise to a sequence

$$0 \longrightarrow \mathcal{H}^q \longrightarrow \coprod_{x \in X^{(0)}} H^q(k(x)) \longrightarrow \coprod_{x \in X^{(1)}} H^{q-1}(k(x)) \longrightarrow \dots$$

$$\dots \longrightarrow \coprod_{x \in X^{(q)}} H^0(k(x)) \longrightarrow 0. \tag{8.2}$$

Bloch and Ogus proved that, under suitable conditions on the cohomology theory, this sequence is exact, thus giving a proof of an analogue of Gersten's conjecture in algebraic K-theory. These conditions on the (co)homology theory led Bloch and Ogus [BO] to introduce their twisted Poincaré duality theory with supports.

The filtration by coniveau plays an important role in several hard conjectures in algebraic geometry, e.g. the Hodge and Tate Conjectures. Unfortunately it is rather mysterious in most cases.

As a first example we consider the Hodge Conjecture as formulated by Grothendieck in [Gr2]. Grothendieck's formulation of the general Hodge conjecture can be stated in terms of the filtration by coniveau as follows:

Conjecture 8.1.1 (Grothendieck) *Let X be a smooth, complex variety of dimension n, and let $0 \leq i \leq 2n$. Then every $\mathbf{Q}$–Hodge substructure V of $H^i(X, \mathbf{Q})$ with level $\leq i-2j$ is contained in $N^jH^i(X, \mathbf{Q})$.*

Here the level of V is defined as $\max\{|p-q| \mid V^{p,q} \neq 0\}$, where the $V^{p,q}$ are the components of the Hodge decomposition of $V \otimes \mathbf{C} = \sum V^{p,q}$.

The classical Hodge Conjecture corresponds to the case $i = 2j$. It says that every element of $H^{2j}(X, \mathbf{Q}) \cap H^{j,j}$ is a rational multiple of the cohomology class of an algebraic cycle on X. In terms of motivic cohomology this conjecture can be translated as follows:

Conjecture 8.1.2 *The image of*

$$r_{\mathcal{D}} : H^{2j}_{\mathcal{M}}(X, \mathbf{Q}(j)) \longrightarrow H^{2j}_{\mathcal{D}}(X, \mathbf{R}(j)) = H^{2j}(X, \mathbf{R}(j)) \cap H^{j,j}(X)$$

is equal to $H^{2j}(X, \mathbf{Q}(j)) \cap H^{j,j}(X)$.

Beilinson gives a variant

$$r_{\mathcal{D}} : H^{2j-1}_{\mathcal{M}}(X, \mathbf{Q}(j)) \longrightarrow H^{2j-1}_{\mathcal{D}}(X, \mathbf{R}(j)) =$$

$$= H^{2j-2}(X, \mathbf{R}(j-1)) \cap H^{j-1,j-1}(X)$$

and conjectures $r_{\mathcal{D}}$ to be surjective. More precisely, one has

Conjecture 8.1.3 (Beilinson IV) *Let X be a smooth proper scheme over a field $k \subset \mathbf{R}$. Then, for $i \leq 2j-1$, the regulator map*

$$r_{\mathcal{D}} : H^i_{\mathcal{M}}(X, \mathbf{Q}(j)) \otimes \mathbf{R} \longrightarrow H^i_{\mathcal{D}}(X \otimes_k \mathbf{R}, \mathbf{R}(j))$$

is an isomorphism.

Actually, for $i = 2j$, Beilinson gives a conjecture implying the usual Hodge conjecture for $X(\mathbf{C})$. We come back to this topic when Beilinson's fifth conjecture will be formulated, cf. Conjecture 8.3.1.

By a result of Suslin which says that for a field F the $H^i_{\mathcal{M}}(F, \mathbf{Q}(j))$ vanish for $i > j$, and using Quillen's spectral sequence (Theorem 4.4.5) and a refinement of Soulé, one has the following theorem, (cf. [Ja1], [Ja2]):

Theorem 8.1.1 *For smooth X, $H^i_{\mathcal{M}}(X, \mathbf{Q}(j))$, $i \leq 2j-1$, has support in codimension $i-j$.*

Now Beilinson's Conjecture IV would imply

Corollary 8.1.1 (Hodge $\mathcal{D}$–Conjecture) *If X is a smooth, proper variety over $\mathbf{C}$ and $i \leq 2j-1$, then $H^i_{\mathcal{D}}(X, \mathbf{R}(j))$ has support in codimension $i-j$.*

Example 8.1.1 Consider the case $i = 2j-1$, X smooth, projective over $\mathbf{C}$. Assuming the truth of the Hodge $\mathcal{D}$–Conjecture, one obtains an explicit description of

$$H^{2j-1}_{\mathcal{D}}(X, \mathbf{R}(j)) = H^{2j-2}(X, \mathbf{R}(j-1)) \cap H^{j-1,j-1}(X)$$

in terms of currents in the following way:

For a family of codimension $j-1$ integral, closed subvarieties $\{Y_\alpha\}$ of X, and $f_\alpha \in \mathbf{C}(Y_\alpha)^*$ with

$$\sum_\alpha \operatorname{div}(f_\alpha) = 0 \text{ and } \partial\bar{\partial} f_\alpha = -\pi i \delta_{\operatorname{div}(f_\alpha)},$$

the $\log|f_\alpha|$ are currents given by

$$(\log|f_\alpha|)(\omega) = -\int_{Y_\alpha \setminus Y_\alpha^{sing}} \log|f_\alpha| \wedge \omega .$$

The elements of $H^{2j-2}(X, \mathbf{R}(j-1)) \cap H^{j-1,j-1}(X)$ are then the classes of currents $\sum_\alpha \log|f_\alpha|$ in $\operatorname{Ker}(\partial\bar{\partial})/(\operatorname{Im}(\partial) + \operatorname{Im}(\bar{\partial}))$.

8.2 Absolute Hodge Cohomology

Beilinson also states a more general conjecture, which applies to arbitrary smooth schemes over **C**. To state this conjecture one has to generalize Deligne cohomology to take into account the weight filtration of the mixed Hodge structure on the cohomology of such schemes. This leads to the so-called **absolute Hodge cohomology**.

The construction roughly goes as follows: Let A be a noetherian subring of $\mathbf{R}$ such that $A \otimes \mathbf{Q}$ is a field. Denote by $\mathcal{H}$ the category of (mixed) A-Hodge structures. This is an abelian A-linear tensor category with internal Hom, written $\mathcal{H}om$ and tensor product $\otimes$. Also, denote by $A(i) \in \mathcal{O}b(\mathcal{H})$, $i \in \mathbf{Z}$, the ith $\otimes$-power of the Tate object $A(1) \in \mathcal{O}b(\mathcal{H})$, and put $M(i) = M \otimes A(i)$ for any $M \in \mathcal{O}b(\mathcal{H})$.The functor

$$\Gamma_{\mathcal{H}} : \mathcal{H} \longrightarrow A - \mathcal{M}od$$

is defined by $\Gamma_{\mathcal{H}}(M) = \mathrm{Hom}_{\mathcal{H}}(A(0), M)$. This is a left-exact functor and one has a canonical identification

$$\mathrm{Hom}_{\mathcal{H}}(M, N) = \Gamma_{\mathcal{H}}\mathcal{H}om(M, N)\,, \ M, N \in \mathcal{O}b(\mathcal{H})\,.$$

Let $\mathrm{D}^{\bullet}(\mathcal{H})$ denote the derived category of $\mathcal{H}$, where $\bullet$ may be any of the usual boundedness conditions $+, -, \mathrm{b}$ or $\emptyset$. Write $\underline{H} : \mathrm{D}^{\bullet}(\mathcal{H}) \longrightarrow \mathcal{H}$ for the standard cohomological functor.

The **absolute Hodge cohomology functor** is defined as the functor

$$\mathrm{R}\Gamma_{\mathcal{H}} : \mathrm{D}^{+}(\mathcal{H}) \longrightarrow \mathrm{D}^{+}(A - \mathcal{M}od)$$

and the **absolute Hodge cohomology** is given by

$$H^{i}_{\mathcal{H}}(\mathcal{F}^{\bullet}) = H^{i}(\mathrm{R}\Gamma_{\mathcal{H}}(\mathcal{F}^{\bullet}))\,.$$

The triple $(M_A, W_{\mathbf{Q}\bullet}, F^{\bullet})$ is built by an object $M \in \mathcal{O}b(\mathcal{H})$ with underlying A-module M_A, weight filtration $W_{\mathbf{Q}\bullet}$ on $M_{\mathbf{Q}} = M_A \otimes \mathbf{Q}$ and Hodge filtration $F^{\bullet}$ on $M_{\mathbf{C}} = M_A \otimes \mathbf{C}$. Also, write $W_{\mathbf{C}\bullet} = W_{\mathbf{Q}\bullet} \otimes \mathbf{C}$ for the induced filtration on $M_{\mathbf{C}}$ and $W_{\bullet}$ for the filtration on the $A \otimes \mathbf{Q}$-mixed Hodge structure $M \otimes \mathbf{Q}$ such that $W_i(M \otimes \mathbf{Q})_{A \otimes \mathbf{Q}} = W_{\mathbf{Q}i}(M)$.

Similarly for complexes of mixed Hodge structures. Next, define a diagram of A-complexes

$$\mathcal{D}_{\mathcal{H}}(\mathcal{F}^\bullet) = \left(\begin{array}{ccccc} & \mathcal{F}^\bullet_{\mathbb{Q}} & & W_{\mathbb{C}0}(\mathcal{F}^\bullet) & \\ \alpha_1 \nearrow & & \nwarrow \beta_1 \ \alpha_2 \nearrow & & \nwarrow \beta_2 \\ \mathcal{F}^\bullet_A & & W_{\mathbb{Q}0}(\mathcal{F}^\bullet) & & (F^0 \cap W_{\mathbb{C}0})(\mathcal{F}^\bullet) \end{array} \right),$$

and let $\alpha = \alpha_1 + \alpha_2$ and $\beta = \beta_1 + \beta_2$. Write

$\tilde{\Gamma}^0_{\mathcal{H}} = \mathcal{F}^\bullet_A \oplus W_{\mathbb{Q}0}(\mathcal{F}^\bullet) \oplus (F^0 \cap W_{\mathbb{C}0})(\mathcal{F}^\bullet)$ and $\tilde{\Gamma}^1_{\mathcal{H}}(\mathcal{F}^\bullet) = \mathcal{F}^\bullet_{\mathbb{Q}} \oplus W_{\mathbb{C}0}(\mathcal{F}^\bullet)$.

Then

$$\alpha - \beta : \tilde{\Gamma}^0_{\mathcal{H}}(\mathcal{F}^\bullet) \longrightarrow \tilde{\Gamma}^1_{\mathcal{H}}(\mathcal{F}^\bullet)$$

with $(\alpha - \beta)(f_A, f_W, f_F) = (\alpha_1 f_A - \beta_1 f_W, \alpha_2 f_W - \beta_2 f_F)$. Let

$$\tilde{\Gamma}_{\mathcal{H}}(\mathcal{F}^\bullet) = \mathrm{Cone}(\alpha - \beta : \tilde{\Gamma}^0_{\mathcal{H}}(\mathcal{F}^\bullet) \longrightarrow \tilde{\Gamma}^1_{\mathcal{H}}(\mathcal{F}^\bullet))[-1], \text{ and}$$

$$\Gamma^0_{\mathcal{H}}(\mathcal{F}^\bullet) = \mathrm{Ker}(\alpha - \beta), \text{ and } \Gamma^1_{\mathcal{H}}(\mathcal{F}^\bullet) = \mathrm{Coker}(\alpha - \beta).$$

This gives a distinguished triangle in $\mathrm{D}(A - \mathcal{M}od)$:

$$\Gamma^0_{\mathcal{H}}(\mathcal{F}^\bullet) \longrightarrow \tilde{\Gamma}_{\mathcal{H}}(\mathcal{F}^\bullet) \longrightarrow \Gamma^1_{\mathcal{H}}(\mathcal{F}^\bullet)[-1] \xrightarrow{[1]} \dots. \tag{8.3}$$

The canonical injection $\Gamma_{\mathcal{H}}(\mathcal{F}^\bullet) \hookrightarrow \tilde{\Gamma}_{\mathcal{H}}(\mathcal{F}^\bullet)$ gives an isomorphism

$$\mathrm{R}\Gamma_{\mathcal{H}}(\mathcal{F}^\bullet) \xrightarrow{\sim} \tilde{\Gamma}_{\mathcal{H}}(\mathcal{F}^\bullet)$$

of $\mathrm{D}^+(A - \mathcal{M}od)$-valued functors on $\mathrm{D}^+(\mathcal{H})$. The canonical filtration $\tau_{\leq\bullet}(\mathcal{F}^\bullet)$ of $\mathcal{F}^\bullet$ induces a filtration $\tilde{\Gamma}_{\mathcal{H}}(\tau_{\leq\bullet}(\mathcal{F}^\bullet))$ of $\tilde{\Gamma}_{\mathcal{H}}(\mathcal{F}^\bullet)$, whose i^{th} graded factor is quasi-isomorphic with $\tilde{\Gamma}_{\mathcal{H}}(\underline{H}^i(\mathcal{F}^\bullet))[-i]$. The corresponding spectral sequence degenerates and one obtains a canonical short exact sequence:

$$0 \longrightarrow \Gamma^1_{\mathcal{H}}(\underline{H}^{i-1}(\mathcal{F}^\bullet)) \longrightarrow H^i\mathrm{R}\Gamma_{\mathcal{H}}(\mathcal{F}^\bullet) \longrightarrow \Gamma^0_{\mathcal{H}}(\underline{H}^i(\mathcal{F}^\bullet) \longrightarrow 0. \tag{8.4}$$

Moreover, the isomorphism $\mathrm{R}\Gamma_{\mathcal{H}} \xrightarrow{\sim} \tilde{\Gamma}_{\mathcal{H}}$ generalizes as follows: For $\mathcal{F}^\bullet \in \mathcal{O}b(\mathrm{D}^+(\mathcal{H}))$ and $\mathcal{G}^\bullet \in \mathcal{O}b(\mathrm{D}^b(\mathcal{H}))$, the arrow

$$\mathrm{RHom}^\bullet(\mathcal{G}^\bullet, \mathcal{F}^\bullet) \longrightarrow \tilde{\Gamma}_{\mathcal{H}}\mathrm{R}\mathcal{H}om^\bullet(\mathcal{G}^\bullet, \mathcal{F}^\bullet)$$

is an isomorphism. This implies, for mixed Hodge structures M and N and $i > 1$, that $\mathrm{Ext}^i_{\mathcal{H}}(M,N) = \mathrm{Ext}^i_A(M_A, N_A)$ and, in particular, if $A = \mathbf{Z}$ or A is a field, that

$$\mathrm{Ext}^i_{\mathcal{H}}(M,N) = 0 \text{ for } i > 1\,. \tag{8.5}$$

This vanishing of the higher Ext will turn up anew in the context of mixed motives where the analogous property is conjectured. For a complex $\mathcal{F}^\bullet \in \mathcal{O}b(\mathrm{D}^+(\mathcal{H}))$ one gets a quasi-isomorphism

$$\mathcal{F}^\bullet \xrightarrow{\sim} \bigoplus_i \underline{H}^i(\mathcal{F}^\bullet)[-i]\,. \tag{8.6}$$

Remark 8.2.1 The whole construction above may be carried through for the abelian subcategory $\mathcal{H}^p$ of polarizable mixed A-Hodge structures, where the mixed Hodge structure $M \in \mathcal{O}b(\mathcal{H})$ is called **polarizable** if the pure weight structures $Gr_i^W(M \otimes \mathbf{Q})$ are polarizable.

A more amenable notion, however, is the one of a (polarizable) **A-Hodge complex**. By this is meant a diagram, denoted by $\mathcal{F}^\bullet$, of the form

$$\begin{array}{ccccc} & \mathcal{F}^{\bullet\prime}_{\mathbf{Q}} & & (\mathcal{F}^{\bullet\prime}_{\mathbf{C}}, W'_{\mathbf{C}\bullet}) & \\ {}^{\alpha_1}\nearrow & & \nwarrow^{\alpha_2}\ {}^{\beta_1}\nearrow & & \nwarrow^{\beta_2} \\ \mathcal{F}^\bullet_A & \xrightarrow[\alpha]{} & (\mathcal{F}^\bullet_{\mathbf{Q}}, W_{\mathbf{Q}\bullet}) & \xrightarrow[\beta]{} & (\mathcal{F}^\bullet_{\mathbf{C}}, W_{\mathbf{C}\bullet}, F^\bullet)\,, \end{array} \tag{8.7}$$

where α_1, α_2, β_1, β_2 are (filtered) quasi-isomorphisms such that

(*i*) $\mathcal{F}^\bullet_A \in \mathcal{O}b(\mathrm{D}^+(A - \mathcal{M}od))$ and $H^\bullet(\mathcal{F}^\bullet_A)$ are complexes of finitely generated A-modules and only finitely many of them are non-zero.

(*ii*) $(\mathcal{F}^\bullet_{\mathbf{Q}}, W_{\mathbf{Q}\bullet})$ is a filtered complex of $A \otimes \mathbf{Q}$–modules and α is a morphism in $\mathrm{D}^\bullet(A - \mathcal{M}od)$ such that $\mathcal{F}^\bullet_A \otimes \mathbf{Q} \xrightarrow{\sim} \mathcal{F}^\bullet_{A\otimes\mathbf{Q}}$ is a quasi-isomorphism. $W_{\mathbf{Q}\bullet}$ is an increasing filtration.

(*iii*) $(\mathcal{F}^\bullet_{\mathbf{C}}, W_{\mathbf{Q}\bullet}, F^\bullet) \in \mathcal{O}b(\mathrm{D}^+(\mathbf{C} - \mathcal{V}ect))$, a bifiltered complex of $\mathbf{C}$–vector spaces and β is a morphism in $\mathrm{D}^+\mathrm{Fil}(A - \mathcal{M}od)$ such that the natural morphism $(\mathcal{F}^\bullet_{\mathbf{Q}} \otimes \mathbf{C}, W_{\mathbf{Q}\bullet} \otimes \mathbf{C}) \xrightarrow{\sim} (\mathcal{F}^\bullet_{\mathbf{C}}, W_{\mathbf{C}\bullet})$ is a filtered quasi-isomorphism. $F^\bullet$ is a decreasing filtration.

(*iv*) The differentials of the complex $\mathrm{Gr}_m^W \mathcal{F}_{\mathbf{C}}^{\bullet}$ are strictly compatible with the filtration $F^{\bullet}$, and $F^{\bullet}$ induces a pure polarizable $A \otimes \mathbf{Q}$–Hodge structure of weight $m+n$ on $H^n(\mathrm{Gr}_m^W \mathcal{F}_{A\otimes\mathbf{Q}}^{\bullet})$ for all $n \in \mathbf{Z}$.

One proves that the filtrations $W_{\mathbf{Q}\bullet}$, $W_{\mathbf{C}\bullet}$ and $F^{\bullet}$ together with the isomorphisms

$$H^{\bullet}(\mathcal{F}_A^{\bullet}) \otimes_A \mathbf{Q} \xrightarrow{\sim} H^{\bullet}(\mathcal{F}_{\mathbf{Q}}^{\bullet}) \text{ and } H^{\bullet}(\mathcal{F}_{\mathbf{Q}}^{\bullet}) \otimes_A \mathbf{C} \xrightarrow{\sim} H^{\bullet}(\mathcal{F}_{\mathbf{C}}^{\bullet})$$

that come from the diagram, define a polarizable A-Hodge structure on $H^{\bullet}(\mathcal{F}_A^{\bullet})$. This polarizable Hodge structure will be denoted by $\underline{H}^{\bullet}(\mathcal{F}^{\bullet}) \in \mathcal{O}b(\mathcal{H}^p)$. A morphism of two Hodge complexes is a morphism of the corresponding diagrams. Let $\mathrm{K}_{\mathcal{H}^p}^{\bullet}$ ($\bullet$ is $+$, $-$, b or $\emptyset$) be the homotopy category of the category of Hodge complexes. Then

$$\underline{H} : \mathrm{K}_{\mathcal{H}^p}^{\bullet} \longrightarrow \mathcal{H}^p$$

is a homological functor. Localizing $\mathrm{K}_{\mathcal{H}^p}^{\bullet}$ by quasi-isomorphisms one finally gets a category $\mathrm{D}_{\mathcal{H}^p}^{\bullet}$. By the cone construction $\mathrm{D}_{\mathcal{H}^p}^{\bullet}$ becomes a triangulated category. One has an obvious exact functor

$$\Phi : \mathrm{D}^b(\mathcal{H}^p) \longrightarrow \mathrm{D}_{\mathcal{H}^p}^b$$

given by $\Phi(\mathcal{F}^{\bullet}) = \tilde{\mathcal{F}}^{\bullet}$ such that $\tilde{\mathcal{F}}_A^{\bullet} = \mathcal{F}_A^{\bullet}$, $\Phi(W_{\mathbf{Q}i})(\tilde{\mathcal{F}}_{\mathbf{Q}}^j) = W_{\mathbf{Q}i+j}(\mathcal{F}_{\mathbf{Q}}^j)$ and $\Phi(F^{\bullet}) = F^{\bullet}$. Beilinson proves:

Theorem 8.2.1 (Beilinson) *The functor* $\Phi : \mathrm{D}^b(\mathcal{H}^p) \longrightarrow \mathrm{D}_{\mathcal{H}^p}^b$ *is an equivalence of triangulated categories.*

Remark 8.2.2 Disregarding polarizability one also has a similar functor $\mathrm{D}^b(\mathcal{H}) \longrightarrow \mathrm{D}_{\mathcal{H}}$ in an obvious way. Beilinson shows that this functor is also an equivalence of triangulated categories.

8.3 Geometric Interpretation

We come to a geometric interpretation of absolute Hodge cohomology and Beilinson's theorem.

Let $\mathcal{S}ch/\mathbf{C}$ denote the category of schemes over $\mathbf{C}$. For an arbitrary $X \in \mathcal{O}b(\mathcal{S}ch/\mathbf{C})$, Deligne [Del] defined a polarizable Hodge complex by replacing X by a smooth simplicial scheme $\tilde{X}_\bullet$, and then compactifying by $\tilde{X}_\bullet \longrightarrow \bar{\tilde{X}}_\bullet$, such that $\tilde{D}_\bullet = \bar{\tilde{X}}_\bullet \backslash \tilde{X}_\bullet$ is a simplicial normal crossings divisor. Deligne's construction then furnishes a cosimplicial polarizable Hodge complex, and by taking the corresponding simple complex, one obtains a polarizable Hodge complex $\underline{\mathrm{R}\Gamma}(X, A) \in \mathcal{O}b(\mathrm{D}^b_{\mathcal{H}^p})$, or, using the same notation and Beilinson's theorem, a complex of polarizable mixed Hodge structures $\underline{\mathrm{R}\Gamma}(X, A) \in \mathcal{O}b(\mathrm{D}^b(\mathcal{H}^p))$. The cohomology of such a complex, $H^\bullet\underline{\mathrm{R}\Gamma}(X, A) \in \mathcal{O}b(\mathcal{H}^p)$, gives back the well known mixed Hodge structure on $H^\bullet(X, A)$ as defined by Deligne. One can also construct a variant with compact supports $\underline{\mathrm{R}\Gamma}_c(X, A) \in \mathcal{O}b(\mathrm{D}^b_{\mathcal{H}^p})$ for $X \in \mathcal{O}b(\mathcal{S}ch_*/\mathbf{C})$, where $\mathcal{S}ch_*/\mathbf{C}$ is the category of schemes over $\mathbf{C}$ and proper morphisms. There is a canonical morphism $\underline{\mathrm{R}\Gamma}_c(X, A) \longrightarrow \underline{\mathrm{R}\Gamma}(X, A)$ which is an isomorphism when X is proper. For a closed subscheme $Y \subset X$ there is a distinguished triangle

$$\ldots \longrightarrow \underline{\mathrm{R}\Gamma}_c(X\backslash Y, A) \longrightarrow \underline{\mathrm{R}\Gamma}_c(X, A) \longrightarrow \underline{\mathrm{R}\Gamma}_c(Y, A) \longrightarrow \ldots .$$

There is also a homological counterpart with compact supports and a Borel-Moore version with all standard functoriality properties. One is led to define the **absolute Hodge (co)homology of a scheme** X over $\mathbf{C}$ as follows:

Definition 8.3.1 *Let X be a scheme over* $\mathbf{C}$. *Then:*

(*i*) *The* **absolute Hodge cohomology** *of X is*

$$H^\bullet_{\mathcal{H}}(X, A(\star)) = H^\bullet_{\mathcal{H}}(\underline{\mathrm{R}\Gamma}(X, A)(\star)) = H^\bullet\mathrm{R}\Gamma_{\mathcal{H}}(\underline{\mathrm{R}\Gamma}(X, A)(\star)) .$$

(*ii*) *The* **absolute Hodge homology** *of X is*

$$H^{\mathcal{H}}_\bullet(X, A(\star)) = H^{BM}_{\mathcal{H}\bullet}(X, A(\star)) =$$
$$= H^\bullet\mathrm{R}\Gamma_{\mathcal{H}}(\mathrm{R}\mathcal{H}om(\underline{\mathrm{R}\Gamma}_c(X, A), A(\star))) = \mathrm{Hom}^\bullet(\underline{\mathrm{R}\Gamma}_c(X, A), A(\star)) .$$

Absolute Hodge (co)homology has several useful properties:

(*i*) $(H^\bullet_{\mathcal{H}}, H^{\mathcal{H}}_\bullet)$ is a twisted Poincaré duality theory with supports.

(*ii*) There is a cycle map and for X smooth, projective, $H^{2i}_{\mathcal{H}}(X, \mathbf{Z}(i))$ coincides with the usual extension of the group of integral $(1,1)$-cocycles and Griffiths's intermediate Jacobian $J^i(X)$.

(*iii*) The formalism of Chern classes and the Chern character in higher K-theory (for $A \supset \mathbf{Q}$) applies. Thus one has a Riemann-Roch theorem and regulator maps from motivic (co)homology to absolute Hodge (co)homology. This reduces to the cases discussed in previous chapters for smooth, projective varieties etc., in particular, for a point one recovers Borel's regulator map.

Assume that $A = \mathbf{Q}$. The canonical short exact sequence (8.4) for X a smooth scheme over $\mathbf{C}$, reads as follows:

$$0 \to \Gamma^1_{\mathcal{H}}(\underline{H}^{i-1}(X, \mathbf{Q}(j)) \longrightarrow H^i_{\mathcal{H}}(X, \mathbf{Q}(j)) \longrightarrow \Gamma^0_{\mathcal{H}}(\underline{H}^i(X, \mathbf{Q}(j)) \to 0\,, \tag{8.8}$$

where $\underline{H}^\bullet(X, \mathbf{Q}(\star)) = \underline{H}^\bullet \underline{\mathrm{R}\Gamma}(X, \mathbf{Q})(\star)$. Using Deligne's mixed Hodge structure, given in terms of the filtrations $(W_\bullet, F^\bullet)$, and the fact that the shifted weight and Hodge filtrations satisfy $W_k(M(j)) = W_{k+2j}(M)$ and $F^m(M(j)) = F^{m+j}(M)$ for an element $M \in \mathcal{O}b(\mathcal{H})$, the short exact sequence (8.8) can be written as

$$\begin{gathered} 0 \longrightarrow W_{2j}H^{i-1}(X, \mathbf{C})/[W_{2j}H^{i-1}(X, \mathbf{Q}(j)) + (F^j \cap W_{2j})H^{i-1}_{DR}(X)] \longrightarrow \\ \longrightarrow H^i_{\mathcal{H}}(X, \mathbf{Q}(j)) \longrightarrow Hg^{i,j}(X) \longrightarrow 0\,, \end{gathered} \tag{8.9}$$

with $Hg^{i,j}(X) = W_{2j}H^i(X, \mathbf{Q}(j)) \cap F^j H^i_{DR}(X)$, the group of (i,j)-Hodge classes or the Hodge theoretic part of absolute Hodge cohomology. For X smooth and projective, one has $Hg^{i,j}(X) = 0$ for $i \neq 2j$, and $Hg^{2j,j}(X) = Hg^j(X)$, the usual Hodge group. The first quotient in the above exact sequence can be considered as a generalization of the intermediate Jacobian.

Both the 'Hodge-theoretic' and the 'Jacobian' parts carry an obvious topology: the 'Hodge-part' is discrete and the 'Jacobian-part' is continuous. Therefore $H^i_{\mathcal{H}}(X, \mathbf{Q}(j))$ has a natural topology. With all this in mind, Beilinson [Be2] states the following conjecture:

Conjecture 8.3.1 (Beilinson V) *For a smooth scheme X over $\mathbf{C}$, the regulator map*

$$r : H^i_{\mathcal{M}}(X, \mathbf{Q}(j)) \longrightarrow H^i_{\mathcal{H}}(X, \mathbf{Q}(j))$$

has dense image.

Remark 8.3.1 For X smooth and projective over $\mathbf{C}$ and $i = 2j$, the conjecture amounts to the classical Hodge Conjecture.

Remark 8.3.2 For X smooth and projective over $\mathbf{C}$ and $i = 2j - 1$, one recovers the example 8.1.1.

Remark 8.3.3 Conjecture V is **false** as it stands, namely U. Jannsen [Ja2] showed that for a smooth variety X over $\mathbf{C}$, the composed map, also called Chern character or regulator map,

$$\begin{aligned} \mathrm{ch}_{2j-1,j} : H^{2j-1}_{\mathcal{M}}(X, \mathbf{Q}(j)) \longrightarrow Hg^{2j-1,j}(X) = \\ = W_{2j}H^{2j-1}(X, \mathbf{Q}(j)) \cap F^j H^{2j-1}_{DR}(X), \quad (8.10) \end{aligned}$$

is in general **not surjective** for $j \geq 2$, (cf. *infra*). Jannsen's result is a corollary of a theorem on the (non)-injectivity of an Abel-Jacobi map related to $\mathrm{ch}_{2j-1,j}$. The non-injectivity of Abel-Jacobi maps in certain situations was demonstrated by D. Mumford [Mu], but Mumford's counter-example involves 'generic cycles' defined over fields of higher transcendence degree. For number fields one expects a completely different situation.

Consequently, one may state the following conjecture:

Conjecture 8.3.2 (Beilinson-Jannsen V′) *Let $X/\mathbf{C}$ be a smooth scheme, that can be defined over a (finite) number field. Then, for all $i, j \in \mathbf{Z}$, the regulator map*

$$r : H^i_{\mathcal{M}}(X, \mathbf{Q}(j)) \longrightarrow H^i_{\mathcal{H}}(X, \mathbf{Q}(j))$$

has dense image.

8.4 Abel-Jacobi Maps

The canonical exact sequence (8.4) admits another interpretation in case $A = \mathbf{Z}$. One sees that

$$\Gamma^1_{\mathcal{H}}(\underline{H}^{i-1}(X, \mathbf{Z}(j)) = \mathrm{Ext}^1_{\mathcal{H}}(\mathbf{Z}, \underline{H}^{i-1}(X, \mathbf{Z}(j)))$$

and

$$\Gamma^0_{\mathcal{H}}(\underline{H}^i(X, \mathbf{Z}(j)) = \mathrm{Hom}_{\mathcal{H}}(\mathbf{Z}, \underline{H}^i(X, \mathbf{Z}(j)),$$

where $\mathbf{Z}$ means the Tate structure $\mathbf{Z}(0) \in Ob(\mathcal{H})$. Thus one obtains the following exact sequence:

$$0 \longrightarrow \mathrm{Ext}^1_{\mathcal{H}}(\mathbf{Z}, \underline{H}^{i-1}(X, \mathbf{Z}(j))) \longrightarrow H^i_{\mathcal{H}}(X, \mathbf{Z}(j)) \longrightarrow$$
$$\longrightarrow \mathrm{Hom}_{\mathcal{H}}(\mathbf{Z}, \underline{H}^i(X, \mathbf{Z}(j))) \longrightarrow 0\,. \quad (8.11)$$

For (mixed) Hodge structures $M \in \mathcal{O}b(\mathcal{H})$ one may write $\Gamma_{\mathcal{H}}(M)$ for $\Gamma^0_{\mathcal{H}}(M)$ and $\mathrm{R}^1\Gamma_{\mathcal{H}}(M)$ for $\Gamma^1_{\mathcal{H}}(M)$. Also, to abbreviate notation, we will use $H^i(X,j)$ for $\underline{H}^i(X, \mathbf{Z}(j))$, considered as a mixed Hodge structure.

Again, the case $i = 2j$ admits a well known interpretation for smooth X as follows: Let $z \in CH^j(X)$ be (the class of) a codimension j cycle on X with support $Z \subset X$. Write $Y = X \backslash Z$, then one has an exact sequence:

$$0 \longrightarrow H^{2j-1}(X,j) \longrightarrow H^{2j-1}(Y,j) \longrightarrow H^{2j}_Z(X,j) \longrightarrow H^{2j}(X,j)\,,$$

where we use that $H_i(Z,j) = 0$, $i > 2\dim(Z)$ for the underlying Poincaré duality theory.

Assume that $z \in A^j(X) = CH^j(X)^0$ is a codimension j cycle homologically equivalent to zero, then by pull-back, z gives rise to an extension $E \in \mathrm{R}^1\Gamma_{\mathcal{H}} H^{2j-1}(X,j) = \mathrm{Ext}^1_{\mathcal{H}}(\mathbf{Z}, H^{2j-1}(X,j))$ defined by the following diagram:

$$\begin{array}{ccccccccc}
0 \to & H^{2j-1}(X,j) & \to & H^{2j-1}(Y,j) & \to & H^{2j}_Z(X,j) & \to & H^{2j}(X,j) & \\
 & \| & & \cup & & \uparrow z & \nearrow 0 & & \\
0 \to & H^{2j-1}(X,j) & \to & E & \to & \mathbf{Z} & \to & 0 &
\end{array} \quad (8.12)$$

where the fundamental class η_z of z is an element of $\Gamma_{\mathcal{H}} H^{2j}_Z(X,j) = \mathrm{Hom}_{\mathcal{H}}(\mathbf{Z}, \underline{H}^{2j}_Z(X, \mathbf{Z}(j)))$. The map

$$\phi_j : A^j(X) \longrightarrow \mathrm{Ext}^1_{\mathcal{H}}(\mathbf{Z}, H^{2j-1}(X,j))$$

is called the **Abel-Jacobi map**. For smooth, projective X over $\mathbf{C}$ one can prove that $\mathrm{Ext}^1_{\mathcal{H}}(\mathbf{Z}, H^{2j-1}(X,j))$ is equal to Griffiths's intermediate Jacobian

$$J^j(X) = H^{2j-1}(X, \mathbf{C}) / H^{2j-1}(X, \mathbf{Z}(j)) \oplus F^j H^{2j-1}(X)\,,$$

and the map $\phi_j : A^j(X) \longrightarrow J^j(X)$ is the classical Abel-Jacobi map.

For a smooth scheme X over $\mathbf{C}$ one has a commutative diagram:

$$\begin{array}{ccccccc} 0 \longrightarrow & A^j(X) & \longrightarrow & CH^j(X) & \longrightarrow & B^j(X) & \longrightarrow 0 \\ & \Big\downarrow{\scriptstyle \phi_j} & & \Big\downarrow{\scriptstyle cl_{\mathcal{H}}} & & \Big\downarrow{\scriptstyle cl_{\mathcal{H}}} & \\ 0 \longrightarrow & \mathrm{R}^1\Gamma_{\mathcal{H}}H^{2j-1}(X,j) & \longrightarrow & H^{2j}_{\mathcal{H}}(X,\mathbf{Z}(j)) & \longrightarrow & \Gamma_{\mathcal{H}}H^{2j}(X,j) & \longrightarrow 0 \end{array} \qquad (8.13)$$

where $cl_{\mathcal{H}}$ denotes the cycle map in absolute Hodge cohomology. $cl_{\mathcal{H}}$ coincides with $cl_{\mathcal{D}}$ in Deligne cohomology. $B^j(X)$ is defined by the exactness of the upper row of the diagram.

From now on we take $A = \mathbf{Q}$ without changing the notation. Returning to the situation of a smooth, proper X with $Z \subset X$ of pure codimension j, and writing $Y = X \backslash Z$, there is a short exact sequence

$$0 \longrightarrow H^{2j-1}(X,j) \longrightarrow H^{2j-1}(Y,j) \longrightarrow H^{2j}_Z(X,j)^0 \longrightarrow 0\,,$$

where we have put $H^{2j}_Z(X,j)^0 = \mathrm{Ker}(H^{2j}_Z(X,j) \longrightarrow H^{2j}(X,j))$. The functor $\Gamma_{\mathcal{H}} = \mathrm{Hom}_{\mathcal{H}}(\mathbf{Q},-)$ gives $\Gamma_{\mathcal{H}}H^{2j-1}(X,j) = 0$ because the Hodge structure $H^{2j-1}(X,j)$ is pure of weight $-1 < 0 =$ weight of $\mathbf{Q} = \mathbf{Q}(0)$, and therefore there is a long exact sequence

$$0 \longrightarrow \Gamma_{\mathcal{H}}H^{2j-1}(Y,j) \longrightarrow \Gamma_{\mathcal{H}}H^{2j}_Z(X,j)^0 \xrightarrow{\ \delta\ } \mathrm{R}^1\Gamma_{\mathcal{H}}H^{2j-1}(X,j) \longrightarrow \dots . \qquad (8.14)$$

Jannsen shows that this sequence can be included in a commutative diagram:

$$\begin{array}{ccccccc} & H^{2j-1}_{\mathcal{M}}(Y,\mathbf{Q}(j)) & \longrightarrow & H^{2j}_{\mathcal{M},Z}(X,\mathbf{Q}(j))^0 & \xrightarrow{\ \ } & H^{2j}_{\mathcal{M}}(X,\mathbf{Q}(j))^0 & \\ & \Big| & & \Big\downarrow\wr & & \Big\downarrow\wr & \\ & {\scriptstyle r}\Big| & & CH^j_Z(X)^0 \otimes \mathbf{Q} & \longrightarrow & CH^j(X)^0 \otimes \mathbf{Q} & \\ & \Big\downarrow & & \Big\downarrow{\scriptstyle cl_{\mathcal{H}}} & & \Big\downarrow{\scriptstyle \phi_j} & \\ 0 \longrightarrow & \Gamma_{\mathcal{H}}H^{2j-1}(Y,j) & \longrightarrow & \Gamma_{\mathcal{H}}H^{2j}_Z(X,j)^0 & \longrightarrow & \mathrm{R}^1\Gamma_{\mathcal{H}}H^{2j-1}(X,j) & \longrightarrow \cdots \end{array} \qquad (8.15)$$

It can be shown that $cl_{\mathcal{H}}$ is an isomorphism in this diagram, and using this fact one deduces the remarkable theorem:

Theorem 8.4.1 (Jannsen) *In the above situation the* **regulator map** *r is surjective if and only if the* **Abel-Jacobi map** *ϕ_j is injective on the subgroup of cycles with support on Z.*

By this theorem and Mumford's result discussed above, Beilinson's original Conjecture V is disproved. The modified Conjecture V′ implies another one, due independently to Bloch and Beilinson:

Conjecture 8.4.1 (Beilinson-Bloch) *For a smooth, proper variety X over $\mathbf{Q}$, the complex Abel-Jacobi map*

$$CH^i(X)^0 \longrightarrow J^i(X \otimes \mathbf{C})$$

is injective up to torsion.

Remark 8.4.1 The conjecture can be stated equally well for arbitrary number fields. The important restriction is the transcendence degree. The same remark applies to what follows.

8.5 The Tate Conjecture

To get a more 'motivic' formulation of statements and conjectures for varieties over $\mathbf{Q}$ (or any number field), and to see at the same time a relation between the Hodge and Tate Conjectures, one would also like to have an ℓ-adic version of (some of) the above statements and conjectures.

The usual ℓ-adic cohomology has the shortcoming that it does not arise from a derived functor formalism and therefore it is not suited for calculations in derived categories. To circumvent this problem Jannsen [Ja3] introduced the so-called **continuous étale cohomology**. This is a derived functor cohomology and its existence is based on the left exactness of the inverse limit.

To define continuous étale cohomology, let X be a scheme and let $(\mathcal{F}_n)_{n\in J} \in \mathcal{O}b(\mathcal{S}(X_{ét})^{\mathbf{N}})$ be an inverse system of sheaves on $X_{ét}$, the small étale site of X. Also let

$$\Gamma = H^0(X,-) : \mathcal{S}(X_{ét}) \longrightarrow \mathcal{A}b$$

be the section functor. Then Jannsen defines the **continuous étale cohomology** $H^i_{cont}(X,\mathcal{F})$ as follows:

Definition 8.5.1 $H^i_{cont}(X,\mathcal{F}) = H^i(X,(\mathcal{F}_n)_{n\in J}) = \mathrm{R}^i(\varprojlim\Gamma)(\mathcal{F}_n)$.

In particular, for the inverse system $(\mathcal{F}_n) = (\mathbf{Z}/\ell^n(j))$ one obtains the **continuous ℓ-adic cohomology** $H^i_{cont}(X,\mathbf{Z}_\ell(j))$. In the 'geometric' case where X is defined over an algebraically closed field these groups coincide with the usual ℓ-adic cohomology groups of X. On the other hand, in the 'arithmetic' case of a field k with absolute Galois group $G_k = \mathrm{Gal}(\bar{k}/k)$, one has

$$H^i_{cont}(\mathrm{Spec}(k),\mathbf{Z}_\ell(j)) = H^i_{cont}(G_k,\mathbf{Z}_\ell(j)),$$

the continuous group cohomology as defined by Tate, (cf. [Ta5]).

An important property of continuous étale cohomology is the existence of a **Hochschild-Serre spectral sequence**:

$$E_2^{p,q} = H^p_{cont}(G_k, H^q(\bar{X},\mathbf{Z}_\ell(j))) \Longrightarrow H^{p+q}_{cont}(X,\mathbf{Z}_\ell(j)) \qquad (8.16)$$

for a variety X defined over the field k.

There is also a cycle map

$$cl_\ell = cl_{cont} : CH^j(X) \longrightarrow H^{2j}_{cont}(X,\mathbf{Z}_\ell(j))$$

for a smooth variety X, with the usual properties.

For a closed subscheme $Z \subset X$ one has cohomology groups with support

$$H^i_{Z,cont}(X,(\mathcal{F}_n)) = \mathrm{R}^i(\varprojlim\Gamma_Z)(\mathcal{F}_n),$$

where $\Gamma_Z(X,\mathcal{F}) = \mathrm{Ker}(\Gamma(X,\mathcal{F}) \longrightarrow \Gamma(X\backslash Z,\mathcal{F}))$, and the usual long exact sequence ... etc. In particular, one has the $H^i_{Z,cont}(X,\mathbf{Z}_\ell(j))$ for a closed subscheme $Z \subset X$.

Let $f : X \longrightarrow \mathrm{Spec}(k)$ be the structure map. Then it is possible to construct a complex with constructible cohomology $\mathrm{R}f^!\mathbf{Z}_\ell(\star) \in \mathcal{O}b(\mathrm{D}^b_c(X,\mathbf{Z}_\ell))$ and to define the **continuous ℓ-adic homology**

$$H_i^{cont}(X,\mathbf{Z}_\ell(j)) = H^{-i}_{cont}(X,\mathrm{R}f^!\mathbf{Z}_\ell(-j)).$$

There exists a corresponding homological Hochschild-Serre spectral sequence which takes the following form:

$$E^2_{p,q} = H^p_{cont}(G_k, H^{\acute{e}t}_{-q}(\bar{X},\mathbf{Z}_\ell(j))) \Longrightarrow H^{cont}_{-p-q}(X,\mathbf{Z}_\ell(j)), \qquad (8.17)$$

where, by definition, $H_{-q}^{\acute{e}t}(\bar{X}, \mathbf{Z}_\ell(j)) = H_{\acute{e}t}^q(\bar{X}, \mathrm{R}\bar{f}^!\mathbf{Z}_\ell(-j))$ for the extension map $\bar{f} : X \times_k \bar{k} \longrightarrow \mathrm{Spec}(\bar{k})$.

If k is a number field or a finite field, then the $(H^\bullet_{cont}, H_\bullet^{cont})$ form a Poincaré duality theory (with supports), taking values in the abelian, $\mathbf{Z}_\ell$-linear tensor category $\mathcal{R}ep_c(G_k, \mathbf{Z}_\ell)$ of finitely generated $\mathbf{Z}_\ell$-modules with continuous G_k-action. For such a module M one can define

$$\Gamma_\ell(M) = \mathrm{Hom}_{G_k}(\mathbf{Z}_\ell, M) = M^{G_k}$$

and

$$\mathrm{R}^1\Gamma_\ell(M) = \mathrm{Ext}^1_{G_k}(\mathbf{Z}_\ell, M)\,, \text{ etc.}$$

and the canonical exact sequence, corresponding to the one in absolute Hodge cohomology, looks as follows:

$$0 \longrightarrow \mathrm{R}^1\Gamma_\ell(H_{\acute{e}t}^{i-1}(\bar{X}, \mathbf{Z}_\ell(j)) \longrightarrow H^i_{cont}(X, \mathbf{Z}_\ell(j)) \xrightarrow{\mathrm{res}} \Gamma_\ell H^i_{\acute{e}t}(\bar{X}, \mathbf{Z}_\ell(j))\,. \tag{8.18}$$

The restriction res need not be surjective, but it becomes so after tensoring with $\mathbf{Q}_\ell$, so one has a short exact sequence

$$0 \longrightarrow \mathrm{R}^1\Gamma_\ell(H_{\acute{e}t}^{i-1}(\bar{X}, \mathbf{Q}_\ell(j))) \longrightarrow H^i_{cont}(X, \mathbf{Q}_\ell(j)) \longrightarrow$$
$$\longrightarrow \Gamma_\ell H^i_{\acute{e}t}(\bar{X}, \mathbf{Q}_\ell(j)) \longrightarrow 0\,. \tag{8.19}$$

Once again, the case $i = 2j$ is of particular importance, namely a now famous and long-standing conjecture of J. Tate [Ta1] states:

Conjecture 8.5.1 (Tate) *For X a smooth, projective variety defined over a number field or a finite field k with $char(k) \neq \ell$, the composed map*

$$CH^j(X)\otimes\mathbf{Q}_\ell \xrightarrow{cl\otimes\mathbf{Q}_\ell} H^{2j}_{cont}(X, \mathbf{Z}_\ell(j))\otimes\mathbf{Q}_\ell \longrightarrow \Gamma_\ell H^{2j}_{\acute{e}t}(\bar{X}, \mathbf{Z}_\ell(j))\otimes\mathbf{Q}_\ell \xrightarrow{\sim}$$
$$\xrightarrow{\sim} H^{2j}_{\acute{e}t}(\bar{X}, \mathbf{Q}_\ell(j))^{G_k} \text{ is surjective.}$$

For smooth X the cycle map

$$cl_\ell : CH^j(X) \longrightarrow H^{2j}_{cont}(X, \mathbf{Z}_\ell(j))$$

and the Hochschild-Serre spectral sequence give an Abel-Jacobi map

$$A^j(X) \xrightarrow{\phi_j} \mathrm{Ext}^1_{G_k}(\mathbf{Z}_\ell, H^{2j-1}(\bar{X}, \mathbf{Z}_\ell(j)) = H^1_{cont}(G_k, H^{2j-1}_{\acute{e}t}(\bar{X}, \mathbf{Z}_\ell(j)))\,. \tag{8.20}$$

Remark 8.5.1 A formal analogy between the Hodge Conjecture, absolute Hodge cohomology and the complex Abel-Jacobi map on the one hand, and the above Tate Conjecture, (continuous) étale cohomology and the ℓ-adic Abel-Jacobi map on the other hand, is now apparent. Actually, to an algebraic variety over a number field one can associate a complex variety with a (mixed) Hodge structure on its cohomology on the one hand, and consider the Galois module structure of the étale cohomology of the variety on the other hand. These two aspects are interrelated by comparison morphisms and they are at the very basis of the theory of motives, to be discussed in the next two sections. From the 'Hodge side' the fundamental motivic structure is furnished by the pure Hodge structure $\mathbf{Q}$ (or, even more important, $\mathbf{Q}(1)$), and its 'Galois companion' is $\mathbf{Q}_\ell$ (or, better, $\mathbf{Q}_\ell(1)$). They are the components of the 1-element of a suitable tensor category. This remark may give some plausibility for the logical coherence of the diagram below.

Both the (classical) Hodge Conjecture and the above Tate Conjecture can now be compared via the following diagram, where the upper part applies to smooth, projective varieties over $\mathbf{C}$, and the lower part applies to smooth, projective varieties over a finitely generated field, in particular a number field (say $\mathbf{Q}$) or a finite field k with $char(k) \neq \ell$:

$$
\begin{array}{ccccccc}
0 \to \mathrm{Ext}^1_{\mathcal{H}}(\mathbf{Q}, H^{2j-1}(X,\mathbf{Q}(j))) & \longrightarrow & H^{2j}_{\mathcal{H}}(X,\mathbf{Q}(j)) & \to & \Gamma_{\mathcal{H}} H^{2j}(X,\mathbf{Q}(j)) & \to 0 \\
\uparrow \phi_j & & \uparrow cl & \nearrow \text{Hodge} & \uparrow & \\
0 \longrightarrow A^j(X)\otimes \mathbf{Q}_{(\ell)} & \longrightarrow & CH^j(X)\otimes \mathbf{Q}_{(\ell)} & \to & B^j(X)\otimes \mathbf{Q}_{(\ell)} & \to 0 \\
\downarrow \phi_j & & \downarrow cl & \searrow \text{Tate} & \downarrow & \\
0 \to \mathrm{Ext}^1_{G_k}(\mathbf{Q}_\ell, H^{2j-1}_{\acute{e}t}(\bar{X},\mathbf{Q}_\ell(j))) & \to & H^{2j}_{cont}(X,\mathbf{Q}_\ell(j)) & \to & \Gamma_\ell H^{2j}_{\acute{e}t}(\bar{X},\mathbf{Q}_\ell(j)) & \to 0
\end{array}
\tag{8.21}
$$

8.6 Absolute Hodge Cycles

The above diagram has no meaning as it stands because the middle row is ambiguoüs. To make the diagram meaningful we assume from now on that X is defined over a field k that is embeddable into $\mathbf{C}$, e.g. k may be a number field with embeddings $\sigma : k \longrightarrow \mathbf{C}$. Then the $H^{\bullet}(X, \mathbf{Q}(\star))$ and the $H^{\bullet}_{\acute{e}t}(\bar{X}, \mathbf{Q}_\ell(\star))$ may be considered as the Betti, resp. étale realizations of the 'motive' $M = h^{\bullet}(X)$. For the moment this is just a symbol, to be made more precise in the next section. It suffices to keep in mind that for smooth, projective varieties there are various cohomology theories, such as the Betti, the de Rham and the étale ones, that have several properties in common, suggesting that they are mere aspects (realizations) of a universal structure, the 'motive', associated to the variety at hand.

So, let $M = h^i(X)$, $i \geq 0$, be the 'motive' attached to the smooth, projective variety X defined over the field k and let $\sigma : k \longrightarrow \mathbf{C}$ denote an embedding of k into the complex numbers $\mathbf{C}$. For such a motive one has the following realizations:

(*i*) $H_\sigma(M) = H^i_\sigma(X) = H^i(X \otimes_{k,\sigma} \mathbf{C}, \mathbf{Q})$, the Betti realization for the embedding $\sigma : k \longrightarrow \mathbf{C}$.

(*ii*) $H_{DR}(M) = H^i_{DR}(X) = H^i(X/k)$, the de Rham realization.

(*iii*) $H_\ell(M) = H^i_\ell(X) = H^i_{\acute{e}t}(X \otimes_k \bar{k}, \mathbf{Q}_\ell)$, the ℓ-adic realization.

These are related by the comparison isomorphisms coming from the canonical ones for the cohomology of the variety σX:

$$I_{\ell,\bar{\sigma}} : H^i(X \otimes_{k,\sigma} \mathbf{C}, \mathbf{Q}) \otimes_{\mathbf{Q}} \mathbf{Q}_\ell \xrightarrow{\sim} H^i_{\acute{e}t}(X \otimes_{k,\sigma} \mathbf{C}, \mathbf{Q}_\ell) \xrightarrow{\sim} \\ \xrightarrow{\sim} H^i_{\acute{e}t}(X \otimes_k \bar{k}, \mathbf{Q}_\ell)$$

and

$$I_{\infty,\sigma} : H^i(\sigma X, \mathbf{Q}) \otimes_{\mathbf{Q}} \mathbf{C} \xrightarrow{\sim} H^i_{DR}(\sigma X/\mathbf{C}) \xrightarrow{\sim} H^i_{DR}(X/k) \otimes_{k,\sigma} \mathbf{C}.$$

Each of the above realizations (cohomology theories) admits a **Tate twist**, denoted by $H_\alpha(M)(m) = H_\alpha(M \otimes \mathbf{Q}(1)^{\otimes m})$, $\alpha \in \{\sigma, DR, \ell\}$, and where $\mathbf{Q}(1)$ is the **Tate motive**. By definition, the various Tate twists are given by:

$$H^i_\sigma(X)(m) = H^i_\sigma(X) \otimes_{\mathbf{Q}} (2\pi i)^m \mathbf{Q} = H^i(X \otimes_{k,\sigma} \mathbf{C}, (2\pi i)^m \mathbf{Q}),$$
$$H^i_{DR}(X)(m) = H^i_{DR}(X),$$
$$H^i_\ell(X)(m) = H^i_{\acute{e}t}(X \otimes_k \bar{k}, \mathbf{Q}_\ell(m)) = \varprojlim_\nu H^i_{\acute{e}t}(X \otimes_k \bar{k}, \mu_{\ell^\nu}(\bar{k})^{\otimes m}) \otimes_{\mathbf{Z}} \mathbf{Q}.$$

The comparison isomorphisms are compatible with the various Tate twists.

To abbreviate, write $\bar{X}$ for $X \otimes_k \bar{k}$ and $H^i_{ét,f}(X)$ for $\prod_\ell H^i_\ell(X)$. Then one has for the Tate twist

$$H^i_{ét,f}(X)(m) = H^i_{ét,f}(X) \otimes_{\mathbf{A}^f} (\mathbf{A}^f(1))^{\otimes m}$$

with $\mathbf{A}^f(1) = (\varprojlim_\nu \mu_\nu(\bar{k})) \otimes_{\mathbf{Z}} \mathbf{Q}$. Also, write

$$H^i_{\mathbf{A}}(X)(m) = H^i_{DR}(X)(m) \times H^i_{ét,f}(X)(m).$$

Then, for all $\sigma : k \longrightarrow \mathbf{C}$ and $\bar{\sigma} : \bar{k} \longrightarrow \mathbf{C}$ with $\bar{\sigma}|k = \sigma$, the product of the comparison isomorphisms induces an isomorphism

$$\alpha_\sigma : H^i_\sigma(X)(m) \otimes (\mathbf{C} \times \mathbf{A}^f) \xrightarrow{\sim} H^i_{\mathbf{A}}(\sigma X)(m),$$

where $H^i_{\mathbf{A}}(\sigma X)(m) = H^i_{DR}(\sigma X)(m) \times H^i_{ét,f}(\sigma X)(m)$. Write σ^* for the product isomorphism

$$\sigma^* : (H^i_{DR}(X)(m) \otimes_{k,\sigma} \mathbf{C}) \times H^i_{ét,f}(X)(m) \xrightarrow{\sim} H^i_{\mathbf{A}}(\sigma X)(m).$$

Then Deligne defines:

Definition 8.6.1 *$z \in H^{2p}_{\mathbf{A}}(X)(p)$ is a* Hodge cycle relative to σ *if*

(i) $\sigma^*(z)$ lies in the rational subspace *$H^{2p}_\sigma(X)(p)$ of $H^{2p}_{\mathbf{A}}(\sigma X)(p)$.*

(ii) The first (i.e. the de Rham) component z_{DR} of z is contained in $F^0 H^{2p}_{DR}(X)(p) = F^p H^{2p}_{DR}(X)$.

Definition 8.6.2 *If z is a Hodge cycle relative to every embedding $\bar{\sigma} : \bar{k} \longrightarrow \mathbf{C}$ then z is called an* absolute Hodge cycle.

Remark 8.6.1 Sometimes one prefers to include 'Betti–components' z_σ in the definition, thus an absolute Hodge cycle will be an element of $\prod_\sigma H^{2p}_\sigma(X)(p) \times H^{2p}_{\mathbf{A}}(X)(p)$ with properties adapted from the above definitions.

Deligne's main result is the following theorem:

Theorem 8.6.1 (Deligne) *If X is an* abelian variety *over an algebraically closed field $\bar{k}$, and z is a Hodge cycle on X relative to one embedding $\bar{\sigma} : \bar{k} \longrightarrow \mathbf{C}$, then z is an absolute Hodge cycle.*

For $\bar{k} = \mathbf{C}$ this means that the classical notion of a Hodge cycle on an abelian variety X, i.e. an element of $H^{2p}(X, \mathbf{Q}) \cap H^{p,p}(X)$, is intrinsic, i.e. does not depend on the map $X \longrightarrow \operatorname{Spec}(\mathbf{C})$.

8.7 Motives

After this introduction on absolute Hodge cycles we give a more rigorous definition of motives for absolute Hodge cycles, also called Deligne motives.

To this end, let $\mathcal{V}_k$ be the category of smooth, projective varieties over k. Assume $X \in \mathcal{O}b(\mathcal{V}_k)$ has pure dimension n. Then denote by $C^p_{AH}(X)$ the rational vector space of absolute Hodge cycles on X. Moreover, for $X, Y \in \mathcal{O}b(\mathcal{V}_k)$, define

$$\mathrm{Hom}^p_{AH}(X,Y) = C^{n+p}_{AH}(X \times Y)\,,\; n = \dim(X)\,.$$

Then one has

$$\begin{aligned}\mathrm{Hom}^p_{AH}(X,Y) &\subset H^{2n+2p}_{\mathbf{A}}(X \times Y)(p+n)\\ &= \bigoplus_{r+s=2n+2p} H^r_{\mathbf{A}}(X) \otimes H^s_{\mathbf{A}}(Y)(p+n)\\ = \bigoplus_{s=r+2p} H^r_{\mathbf{A}}(X)^\vee \otimes H^s_{\mathbf{A}}(Y)(p) &= \bigoplus_r \mathrm{Hom}(H^r_{\mathbf{A}}(X), H^s_{\mathbf{A}}(Y)(p))\end{aligned}$$

by Künneth and Poincaré duality. An element $\alpha \in \mathrm{Hom}^p_{AH}(X,Y)$ gives rise to families of homomorphisms $(\alpha^r_\sigma, \alpha^r_{DR}, \alpha^r_\ell)$:

$$\begin{cases} \alpha^r_\sigma : & H^r_\sigma(X) \longrightarrow H^{r+2p}_\sigma(Y)(p)\,,\text{ for each } \sigma : \bar{k} \longrightarrow \mathbf{C}\,;\\ \alpha^r_{DR} : & H^r_{DR}(X) \longrightarrow H^{r+2p}_{DR}(Y)(p)\,;\\ \alpha^r_\ell : & H^r_\ell(X) \longrightarrow H^{r+2p}_\ell(Y)(p)\,,\text{ for each prime } \ell\,, \end{cases}$$

corresponding to the comparison isomorphisms. The α^r_{DR} are compatible with the Hodge filtrations and the α^r_ℓ are G_k-invariant. Conversely, such a family defines a unique $\alpha \in \mathrm{Hom}^p_{AH}(X,Y)$.

Next, define the category $\mathcal{CV}_k$ of AH-correspondences as follows: The objects are symbols $h(X)$, one for each $X \in \mathcal{O}b(\mathcal{V}_k)$, and the class of morphisms between to such objects is

$$\mathrm{Hom}(h(X), h(Y)) = \mathrm{Hom}_{\mathcal{CV}_k}(h(X), h(Y)) = \mathrm{Hom}^0_{AH}(X,Y)$$

for $X, Y \in \mathcal{O}b(\mathcal{V}_k)$. The map $\mathrm{Hom}_{\mathcal{V}_k}(Y,X) \longrightarrow \mathrm{Hom}(h(X), h(Y))$ sending a morphism in $\mathcal{V}_k$ to the cohomology class of its graph, makes

$$h : \mathcal{V}_k \longrightarrow \mathcal{CV}_k$$

into a contravariant functor of categories. One has

$$h(X \coprod Y) = h(X) \oplus h(Y)$$

and $\mathcal{CV}_k$ becomes a $\mathbf{Q}$–linear tensor category by defining

$$h(X) \otimes h(Y) = h(X \times Y).$$

The associativity and commutativity constraints are the obvious ones, and the identity object is $1 = h(\mathrm{Spec}(k))$.

Now let $\mathcal{M}_k^{\bullet,+}$ be the pseudo-abelian or Karoubian envelope of $\mathcal{CV}_k$. The objects of $\mathcal{M}_k^{\bullet,+}$ are pairs $(h(X), p)$ where $h(X) \in Ob(\mathcal{CV}_k)$ and p is a projector in $h(X)$, i.e. a $\mathcal{CV}_k$-morphism $p : h(X) \longrightarrow h(X)$ such that $p^2 = p$. A $\mathcal{M}_k^{\bullet,+}$-morphism from $(h(X), p)$ to $(h(Y), q)$ is a $\mathcal{CV}_k$-morphism $f : h(X) \longrightarrow h(Y)$ such that $f \circ p = q \circ f$ modulo those f such that $f \circ p = q \circ f = 0$. Composition of morphisms in $\mathcal{M}_k^{\bullet,+}$ is induced by composition of morphisms in $\mathcal{CV}_k$, the identity morphism of $(h(X), p)$ is p and the sum of two objects $(h(X), p)$ and $(h(Y), q)$ is defined as $(h(X) \oplus h(Y), p \oplus q)$. $\mathcal{M}_k^{\bullet,+}$ is pseudo-abelian now means that every projector has a kernel. The above construction is universal in the sense that for every pseudo-abelian category $\mathcal{C}$ and for every additive functor $\psi : \mathcal{CV}_k \longrightarrow \mathcal{C}$, there exists an additive functor

$$\psi' : \mathcal{M}_k^{\bullet,+} \longrightarrow \mathcal{C}$$

which is unique up to isomorphism and which makes the following diagram

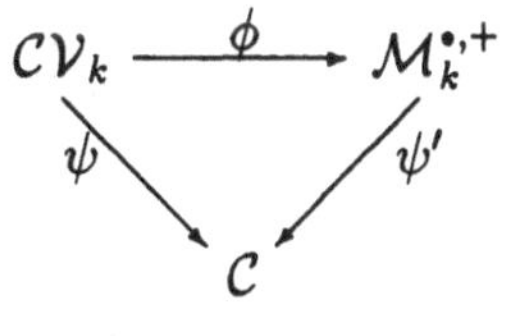

commutative up to isomorphism. The functor ϕ is defined by

$$\phi(h(X)) = (h(X), id_{h(X)}) \in Ob(\mathcal{M}_k^{\bullet,+}),$$

where $id_{h(X)} \in C_{AH}^{\dim(X)}(X \times X)$ is the class of the diagonal $\Delta \subset X \times X$. ϕ is an additive, fully faithful functor and the pair $(\phi, \mathcal{M}_k^{\bullet,+})$ is unique up to equivalence of categories. If p is a projector of the object $h(X)$, then $id_{h(X)} - p$ is also a projector of $h(X)$ and one has in a canonical way

$$h(X) = \mathrm{Ker}(p) \oplus \mathrm{Ker}(id_{h(X)} - p)\,.$$

More generally, if $id_{h(X)}$ decomposes as a sum of mutually orthogonal projectors,

$$id_{h(X)} = \sum_{i=1}^{n} p_i\,,$$

then $h(X) = (h(X), id_{h(X)})$ decomposes as

$$h(X) = \bigoplus_{i=1}^{n} (h(X), p_i) = \bigoplus_{i=1}^{n} h(X)^i\,.$$

In case ψ is fully faithful such that every object of $\mathcal{C}$ is a direct factor of an object in the image of ψ, the functor ψ' is an equivalence of categories. For details we refer to [Ka].

The category $\mathcal{M}_k^{\bullet +}$ is called the **false category of effective motives**.

The projections $p^r : \bigoplus_i H^i(X) \longrightarrow H^r(X)$ define projectors of $h(X)$ and therefore, give rise to a decomposition $h(X) = \bigoplus_r h^r(X)$ which extends uniquely to a $\otimes$-grading on $\mathcal{M}_k^{\bullet +}$ such that $h(X)^r = h^r(X)$. One defines the **Lefschetz motive** L by $L = h^2(\mathbf{P}^1)$. L is homogeneous of degree 2 and $h(\mathbf{P}^1) = 1 \oplus L$. One shows that

$$h(\mathbf{P}^n) = 1 \oplus L \oplus \cdots \oplus L^{\otimes n}\,.$$

The functor $\otimes L$, $V \longmapsto V \otimes L$, is fully faithful and allows one to invert L. The **false category of motives** $\mathcal{M}_k^{\bullet}$ is defined as follows:

(*i*) An object of $\mathcal{M}_k^{\bullet}$ is a pair (M, m) with $M \in \mathcal{O}b(\mathcal{M}_k^{\bullet +})$, $m \in \mathbf{Z}$;

(*ii*) $\mathrm{Hom}_{\mathcal{M}_k^{\bullet}}((M,m),(N,n)) =$

$$= \varinjlim \mathrm{Hom}_{\mathcal{M}_k^{\bullet +}}(M \otimes L^{\otimes(k-m)}, N \otimes L^{\otimes(k-n)})\,,$$

$k \geq m, n$, with evident transition morphisms.

(*iii*) Composition of morphisms is induced by the one in $\mathcal{M}_k^{\bullet +}$.

In short, one has $\mathcal{M}_k^{\bullet} = \mathcal{M}_k^{\bullet +}[L^{-1}]$.

Thus any false motive $M' = (M, m)$ can be written as $M' = M(m) = M \otimes T^{\otimes m}$, where M is an effective motive and $T = L^{-1}$

is called the **Tate motive**, also written $\mathbf{Q}(1)$. In particular, one can write $T = (1,1)$ and $L = (1,-1)$. The tensor product on $\mathcal{M}_k^{\bullet}$ becomes

$$(M,m) \otimes (N,n) = (M \otimes N, m+n)$$

and the $\otimes$-grading on $\mathcal{M}_k^{\bullet+}$ extends to $\mathcal{M}_k^{\bullet}$ by $M(m)^r = M^{r-2m}$.

For $X, Y \in \mathcal{O}b(\mathcal{V}_k)$ of pure dimension m and n, respectively, one has

$$\operatorname{Hom}(h(X), h(Y)) = C^m_{AH}(X \times Y) = C^m_{AH}(Y \times X) =$$
$$= \operatorname{Hom}(h(Y), h(X)(m-n)) = \operatorname{Hom}(h(Y)(n), h(X)(m)),$$

and this suggests the definition of the **dual** $h(X)^\vee = h(X)(m)$ such that $h(X)^\vee \otimes h(X) \longrightarrow 1$. The functor $^\vee$ extends to a fully faithful functor $^\vee : \mathcal{M}_k^{\bullet} \longrightarrow \mathcal{M}_k^{\bullet}$, thus giving rise to an internal Hom, which is usually written $\mathcal{H}om(M,N) = M^\vee \otimes N \in \mathcal{O}b(\mathcal{M}_k^{\bullet})$. Thus $M^\vee = \mathcal{H}om(M,1)$. One has

$$\operatorname{Hom}(M,N) = \operatorname{Hom}(1, M^\vee \otimes N) =$$
$$= \operatorname{Hom}(1 \otimes M, N) = \operatorname{Hom}(1, \mathcal{H}om(M,N)).$$

$\mathcal{M}_k^{\bullet}$ becomes a rigid abelian tensor category[1] with $\operatorname{End}(1) = \mathbf{Q}$. The evaluation morphism $M^\vee \otimes M \longrightarrow 1$ leads to the map

$$\operatorname{End}(M) = \operatorname{Hom}(1, M^\vee \otimes M) \longrightarrow \operatorname{End}(1) = \mathbf{Q}$$

and one calls this morphism the **trace morphism**

$$\operatorname{Tr}_M : \operatorname{End}(M) \longrightarrow \operatorname{End}(1) = \mathbf{Q},$$

in particular, $\operatorname{Tr}_M(id_M)$ is called the **rank** of M, written $\operatorname{rk}(M)$. For $X \in \mathcal{O}b(\mathcal{V}_k)$ one finds

$$\operatorname{rk}(h(X)) = \sum_{r=0}^{2\dim(X)} (-1)^r \dim(H^r(X)),$$

the Euler-Poincaré characteristic. This may be negative. Because one would like the category of motives to be a (neutral) **tannakian category**, i.e. a category equivalent to the category of (finite dimensional)

[1] For definitions and properties of (rigid) tensor categories, cf. [DMOS] or [Sa].

representations of some group scheme, the rank of a motive should be the rank of a representation, i.e. the trace of its identity morphism, and one is led to modify the commutativity constraint on the tensor product in $\mathcal{M}_k^\bullet$. Let $\dot\psi : M \otimes N \xrightarrow{\sim} N \otimes M$, $\dot\psi = \oplus \dot\psi^{r,s}$, $\dot\psi^{r,s} : M^r \otimes N^s \xrightarrow{\sim} N^s \otimes M^r$ be the commutativity constraint on $\mathcal{M}_k^\bullet$. Modify this constraint as follows:

$$\psi : M \otimes N \xrightarrow{\sim} N \otimes M\,,\ \psi = \oplus \psi^{r,s}\,,\ \psi^{r,s} = (-1)^{rs} \dot\psi^{r,s}\,.$$

This leads to the conclusive definition of a motive for absolute Hodge cycles.

Definition 8.7.1 *The category $\mathcal{M}_k^\bullet$ with this new commutativity constraint ψ is called the (true) category $\mathcal{M}_k$ of* Deligne motives.

Theorem 8.7.1 *$\mathcal{M}_k$ is a semi-simple tannakian category over* $\mathbf{Q}$. *For every $M \in \mathcal{O}b(\mathcal{M}_k)$,* End$(M)$ *is a semi-simple* $\mathbf{Q}$*–algebra.*

We end this section with a few examples of the decomposition of the motive $h(X)$ of some specific varieties X. These examples are taken from [DMOS].

Example 8.7.1 Let X be a curve with Jacobian $J = Jac(X)$, then

$$h(X) = 1 \oplus h^1(J) \oplus L\,.$$

Example 8.7.2 More generally, for an irreducible variety X of dimension n the motive $h(X)$ decomposes as

$$h(X) = 1 \oplus h^1(X) \oplus V \oplus (A \otimes L^{\otimes(n-1)}) \oplus L^{\otimes n}\,,$$

where $h^1(X)$ (resp. A) corresponds to the Picard (resp. Albanese) variety of X.

Example 8.7.3 Let X be a unirational variety of dimension $n \leq 3$, then:

(*i*) If $n = 1$, $h(X) = 1 \oplus L$;

(*ii*) If $n = 2$, $h(X) = 1 \oplus rL \oplus L^2$, for some $r \in \mathbf{N}$;

(*iii*) If $n = 3$, $h(X) = 1 \oplus rL \oplus h^1(A)(-1) \oplus rL^2 \oplus L^3$, for some $r \in \mathbf{N}$.

Example 8.7.4 Let X_d^n be the Fermat hypersurface of dimension n and degree d:

$$X_d^n : T_0^d + T_1^d + \cdots + T_{n+1}^d = 0\,,$$

then

$$h(X_d^n) \oplus d\,h(\mathbf{P}^n) = h^n(X_d^{n-1} \times X_d^1)^{\mu_d} \oplus (d-1)h^{n-2}(X_d^{n-2})(-1)\,,$$

where μ_d, the group of dth roots of 1, acts on $X_d^{n-1} \times X_d^1$ according to

$$\zeta(t_0 : t_1 : \ldots : t_n; s_0 : s_1 : s_2) = (t_0 : \ldots : t_{n-1} : \zeta t_n; s_0 : s_1 : \zeta s_2)\,,$$

$\zeta \in \mu_d$.
Here, the superscript μ_d means that one takes $(h^n(X_d^{n-1} \times X_d^1), p)$ with projector $p = \#(\mu_d)^{-1} \sum_\zeta \zeta$.

For details and applications of tannakian categories we refer to [DMOS], [Kl], [Man], [Sa] and [Sch].

8.8 Grothendieck's Conjectures

The last theorem of the previous section is actually the motivation of the rather complicated definition of a motive. Already in the late 1960's Grothendieck suggested the existence of a universal cohomology theory embodied in a conjectural tannakian category, the category of motives. He tried to formulate a nice theory of motives over an arbitrary field and used another more general notion of morphism between the $h(X)$'s, namely more general correspondences. His main goal, the tannakian character of the category of motives, got stuck on the so-called Standard Conjectures on Algebraic Cycles. There are two of them, one of Lefschetz type, and the other of Hodge type. The one of Lefschetz type can be stated in various equivalent forms. Let X be a smooth, projective variety of dimension n over an algebraically closed field k of arbitrary characteristic, let H be a cohomology functor taking values in the category of (augmented, graded commutative) algebras over a field of characteristic zero. There is a cycle map $cl : Z^i(X) \longrightarrow H^{2i}(X)$ [2] and we write

$$C^i(X) = \mathrm{Im}(cl \otimes \mathbf{Q} : Z^i(X) \otimes \mathbf{Q} \longrightarrow H^{2i}(X))$$

[2] We assume the choice of an 'orientation' to dispose of twists.

for the algebraic cohomology classes. Also let $\xi \in H^2(X)$ be the class of a hyperplane section. Then one has a homomorphism

$$\cup\xi^{n-i} : H^i(X) \longrightarrow H^{2n-i}(X)\,,\ i \leq n\,.$$

The Hard Lefschetz Conjecture says that this is an isomorphism for all characteristics. For X over $\mathbf{C}$ this was proved by Lefschetz for Betti cohomology in 1924, and for X over an algebraically closed k Hard Lefschetz was proved by Deligne [De3] for ℓ-adic cohomology in 1979. For $i = 2j$ one has the commutative diagram

$$\begin{array}{ccc} H^{2j}(X) & \xrightarrow{\xi^{n-2j}} & H^{2n-2j}(X) \\ \uparrow & & \uparrow \\ C^j(X) & \longrightarrow & C^{n-j}(X) \end{array}$$

Conjecture 8.8.1 (Grothendieck's Conjecture $A(X)$)

(a) *Hard Lefschetz always holds.*

(b) *If $i = 2j$, ξ^{n-2j} in the diagram induces an isomorphism*

$$C^j(X) \xrightarrow{\sim} C^{n-j}(X)\,.$$

Let $P^i(X)$ be the primitive cohomology, i.e.

$$P^i(X) = \{x \in H^i(X) | \xi^{n-i+1} \cdot x = 0\}\,,\ i \leq n\,.$$

Then any $x \in H^i(X)$ admits the unique decomposition

$$x = \sum_{j \geq \max(i-n,0)} \xi^j \cdot x_j\,,\ \text{with } x_j \in P^{i-2j}(X)\,.$$

This justifies the definition of an operator Λ by

$$\Lambda x = \sum_{j \geq \max(i-n,1)} \xi^{j-1} \cdot x_j\,.$$

This Λ makes the following diagram commutative

$$\begin{array}{ccc} H^i(X) & \xrightarrow{\sim} & H^{2n-i}(X) \\ \Lambda \downarrow & & \downarrow \xi \\ H^{i-2}(X) & \xrightarrow{\sim} & H^{2n-i+2}(X)\,, \end{array}$$

and one has, in particular, that $\Lambda \circ \xi = id$. The Λ's (for varying i) lead to a cohomological correspondence

$$\Lambda_X \in \mathrm{Hom}(H(X), H(X)) \simeq H(X \times X)$$

and one can state

Conjecture 8.8.2 (Grothendieck's $B(X)$) *Λ_X is algebraic.*

It is not very difficult to prove that the truth of $A(X)$ for all X implies the truth of $B(X)$ and *vice versa*, so $A(X)$ and $B(X)$ are equivalent. $B(X)$ is more tractable than $A(X)$ and it is true for projective spaces and the usual rational varieties with cellular decomposition, curves (trivial), surfaces and abelian varieties (Lieberman). It is stable for products, hyperplane sections and specialization (with possible change of characteristic). $B(X)$ (or $A(X)$) implies another conjecture $C(X)$:

Conjecture 8.8.3 (Grothendieck's $C(X)$) *The Künneth components*

$$\pi^i_X \in H^{2n-i} \otimes H^i(X) \text{ of the diagonal } \Delta_X \in C^n(X \times X),$$

are algebraic.

Remark 8.8.1 For $k = \bar{\mathbf{F}}_q$ N. Katz and W. Messing were able to prove $C(X)$, (cf. [KaM]). They used Deligne's results on the Weil Conjectures and ℓ-adic Hard Lefschetz.

To state Grothendieck's conjecture of Hodge type, let

$$C^j_{Pr}(X) = P^{2j}(X) \cap C^j(X).$$

Then there is a $\mathbf{Q}$–valued symmetric bilinear form on $C^j_{Pr}(X)$ given by

$$(x, y) \longmapsto (-1)^j \mathrm{Tr}_X(x \cdot y \cdot \xi^{n-2j}), \ 0 \leq 2j \leq n,$$

where Tr_X is the isomorphism of $H^{2n}(X)$ with the field of characteristic zero, used in the very definition of the cohomology functor H, (cf. *infra*). The conjecture is the following:

Conjecture 8.8.4 (Grothendieck's $Hdg(X)$) *The above symmetric bilinear form is positive definite.*

Remark 8.8.2 In characteristic zero this is true by Hodge theory.

As a last conjecture related to the Standard Conjectures, one has:

Conjecture 8.8.5 ($D(X)$) *Numerical and homological equivalence coincide for algebraic cycles.*

The interrelations between all these conjectures can be summarized in the following

Theorem 8.8.1 (Grothendieck) *The following equivalences hold:*

$$A(X) + Hdg(X) \iff B(X) + Hdg(X) \iff D(X) + Hdg(X),\ \forall X\,.$$

These conjectures remain unproved. Their truth would imply the truth of the Weil Conjectures for varieties over finite fields, proved by Deligne from another point of view. Also, they are at the basis of a much more general theory of motives, i.e. over any field and using the intrinsically defined numerical equivalence for correspondences. This general category of motives $\mathcal{M}_k$ would be (neutral) tannakian if one accepts the Standard Conjectures. This means that $\mathcal{M}_k$ is equivalent to the category $\mathcal{R}ep(G)$ of (finite dimensional) representations of some (huge) affine pro-algebraic Galois group G. The notion of a tannakian category seems to be very basic and there are several interesting examples such as the category of variations of Hodge structures, and the category of certain crystals ... etc.

8.9 Motives and Cohomology

All suitable cohomology theories take their values in triples of the form $(\mathcal{C}, w, T_{\mathcal{C}})$, where $\mathcal{C}$ is a **Q**–linear rigid [3] pseudo-abelian tensor category

[3]This means that $\mathcal{C}$ has an internal Hom with the properties stated in the section on motives and that every object X of $\mathcal{C}$ is reflexive, i.e. isomorphic to its bidual $X^{\vee\vee}$.

(such as the graded $\mathbf{Q}$–, k– or $\mathbf{Q}_\ell$–vector spaces) with a tensor grading (by $\mathbf{Z}$) w of $id_{\mathcal{C}}$, and with an invertible object $T_{\mathcal{C}}$ of degree -2 for w. A cohomology theory is then given by a triple (H, γ, Tr) with H a contravariant functor from $\mathcal{V}_k$ to $\mathcal{C}^+ \subset \mathcal{C}$, the full subcategory of $\mathcal{C}$ of objects of non-negative degree such that $H(X)$ is a unitary algebra of $\mathcal{C}^+$ and $\bigoplus \Gamma(H^{2i}(X)(i)))$ is a graded $\mathbf{Q}$–algebra, where $\Gamma(V) = \mathrm{Hom}(1, V)$. γ is a family $\{\gamma^i\}_{i \in \mathbf{N}}$ of natural $\mathbf{Z}$–linear transformations

$$\gamma_X^i : Z^i(X)/\sim \longrightarrow \Gamma(H^{2i}(X)(i)) = \mathrm{Hom}(1, H^{2i}(X)(i))$$

for an adequate equivalence relation $\sim$ on the codimension i cycles of X. Here an equivalence relation is called adequate if its residue classes form a ring $A(X)$ under intersection product for each $X \in \mathcal{O}b(\mathcal{V}_k)$. Further the operations on cycles corresponding to a morphism $\phi : Y \longrightarrow X$ should induce a multiplicative operator

$$\phi^* : A(X) \longrightarrow A(Y)$$

and a linear operator

$$\phi_* : A^p(Y) \longrightarrow A^{p+(\dim(X)-\dim(Y))}(X),$$

which satisfy the usual projection formula

$$\phi_*(\phi^* Z \cdot W) = Z \cdot \phi_* W$$

whenever the terms are defined. For example, whenever there exists a cycle map

$$cl_X : Z^i(X) \longrightarrow \Gamma(H^{2i}(X)(i)),\ i \geq 0$$

such that $cl_X(x)$ is independent of the k-rational point $x \in X(k)$, one proves that rational equivalence is the finest, and numerical equivalence is the coarsest adequate equivalence relation.

Tr is a family of morphisms $\mathrm{Tr}_X : H(X)(\dim(X)) \longrightarrow 1$ or if one wants $\mathrm{Tr}_X : H^{2\dim(X)}(X)(\dim(X)) \longrightarrow 1$. The triple (H, γ, Tr) is subject to a number of axioms of a certain compatibility nature and a form of Poincaré duality, (cf. [Sa]). One defines morphisms between such triples in an obvious way.

In this way one may construct the category of cohomology theories over the field k, $\mathcal{TC}(k, (\mathcal{C}, w, T_{\mathcal{C}}))$. Then, at least for the Betti-, the de

Rham- and the ℓ-adic, and hopefully also the crystalline, cohomology theories over a field k such that k and $\bar{k}$ admit compatible embeddings into $\mathbf{C}$, one has a factorization:

$$\begin{array}{ccc} \mathcal{V}_k & \xrightarrow{\quad h \quad} & \mathcal{M}_k \\ & {\scriptstyle H}\searrow \quad \swarrow{\scriptstyle \omega_H} & \\ & \mathcal{C} & \end{array}$$

where h means the construction of the category $\mathcal{M}_k$ starting with the functor h, followed by taking the Karoubian envelope ... etc. With this notation one has $h(X) = (h(X), id_{h(X)})$, which may decompose according to a decomposition of $id_{h(X)}$ into a sum of pairwise orthogonal projectors p_i. Any (connected) smooth, projective k-scheme X admits two special projectors $p_0 = p_X$ and $p_n = q_X$. These arise as follows: Take a closed point x of X of degree $d(x) = [k(x) : k]$. Then the element

$$e_X = d(x)^{-1}\, x \in C^n(X) = \mathbf{Q} \otimes Z^n(X)/\sim$$

is independent of the point x and gives an isomorphism $C^n(X) \xrightarrow{\sim} \mathbf{Q}$. Then there are two endomorphisms p_X and q_X of $h(X)$, defined by

$$p_X = \pi_1^*(e_X) \text{ and } q_X = \pi_2^*(e_X) \in C^n(X \times X)\,,$$

which can easily be shown to be projectors. Of course, the π_i's are the projections $X \times X \longrightarrow X$ on the respective components, $i = 1, 2$. One has

$$(X, p_X) \simeq 1 \text{ and } (X, q_X) \simeq L^{\otimes n}\,,$$

giving the canonical decomposition of the motive of the connected k-scheme of dimension $n \neq 0$:

$$h(X) = 1 \oplus h^+(X) \oplus L^{\otimes n}\,,$$

with $h^+(X) = (h(X), id_{h(X)} - p_X - q_X)$. $h^+(X)$ will contain information on the 'moduli' of X, such as its Picard and Albanese varieties.

The functor ω_H is called the realization functor. It forms half of a couple (ω_H, ξ_H), with

$$\xi_H : \omega_H(T) \xrightarrow{\sim} T_C$$

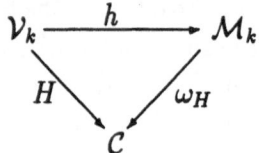

an isomorphism, and ω_H is a rigid tensor functor. The correspondence

$$(H, \gamma, \mathrm{Tr}) \longmapsto (\omega_H, \xi_H)$$

defines an equivalence between $\mathcal{TC}(k, (\mathcal{C}, w, T_{\mathcal{C}}))$ and the category of such couples transforming effective motives to objects of $\mathcal{C}^+$. Such an assertion should hold in much greater generality, i.e. without the restrictions on k, on the choice of absolute Hodge correspondences as morphisms in $\mathcal{CV}_k$ and on the cohomology theories. For more precise statements we refer to [Sa].

In short, one can say, following Manin, that a motive may be considered as a 'twisted direct summand' of a variety, representing at the same time the 'universal cohomology' of the variety.

Example 8.9.1 Let k be a field of characteristic zero admitting embeddings $\sigma : k \longrightarrow \mathbf{C}$ and $\bar{\sigma} : \bar{k} \longrightarrow \mathbf{C}$ such that $\bar{\sigma}|k = \sigma$. $\mathcal{C}$ denotes the category of triples of the form $(\prod_\sigma V_\sigma, V_k, \prod_\ell V_\ell)$, where V_σ, V_k and V_ℓ are (finite dimensional) graded $\mathbf{Q}$–, k– and $\mathbf{Q}_\ell$–vector spaces, respectively. The V_ℓ are G_k-modules and the V_σ, V_k and V_ℓ are related by comparison isomorphisms respecting the G_k-action. One has

$$T_{\mathcal{C}} = (\prod_\sigma T_\sigma, T_k, \prod_\ell T_\ell),$$

where the respective T's are the usual Tate objects as defined before. The 1-object is given by the triple $(\prod_\sigma \mathbf{Q}, k, \prod_\ell \mathbf{Q}_\ell)$ with comparison isomorphisms induced by

$$\mathbf{Q} \longrightarrow \mathbf{C} \stackrel{\sigma}{\longleftarrow} k \text{ and } \mathbf{Q} \longrightarrow \mathbf{Q}_\ell .$$

Let $H : \mathcal{V}_k \longrightarrow \mathcal{C}$ be given by

$$H(X) = \prod_\sigma H_\sigma \times H_{DR}(X) \times \prod_\ell H_\ell(X) = \prod_\sigma H_\sigma(X) \times H_{\mathbf{A}}(X).$$

For the γ_X^i one may take

$$\gamma_X^i = cl_{AH} : CH^i(X) \longrightarrow \Gamma_{AH}(H^{2i}(X)(i)),$$

the absolute Hodge cycles in $H^{2i}(X)(i)$, and $\gamma_X^i = (\prod_\sigma cl_\sigma, cl_k, \prod_\ell cl_\ell)$, where

$$\begin{cases} cl_\sigma : CH^i(X) \xrightarrow{\sigma^*} CH^i(\sigma X) \longrightarrow H_\sigma^{2i}(X)(i)\,, \\ cl_k : CH^i(X) \longrightarrow H_{DR}^{2i}(X)\,, \text{ and} \\ cl_\ell : CH^i(X) \longrightarrow H_\ell^{2i}(X)(i)\,, \end{cases}$$

with the usual identities induced by the comparison isomorphisms.

One obtains the following picture, where we use the obvious notation $H_\sigma^{i,i}(X) = F^i H^{2i}(\sigma X, \mathbf{C})$, and $\Gamma_{alg}(H^{2i}(X)(i)) = \mathrm{Im}\gamma_X^i \otimes 1$, where $1 = (\mathbf{Q}, k, \mathbf{Q}_\ell)$:

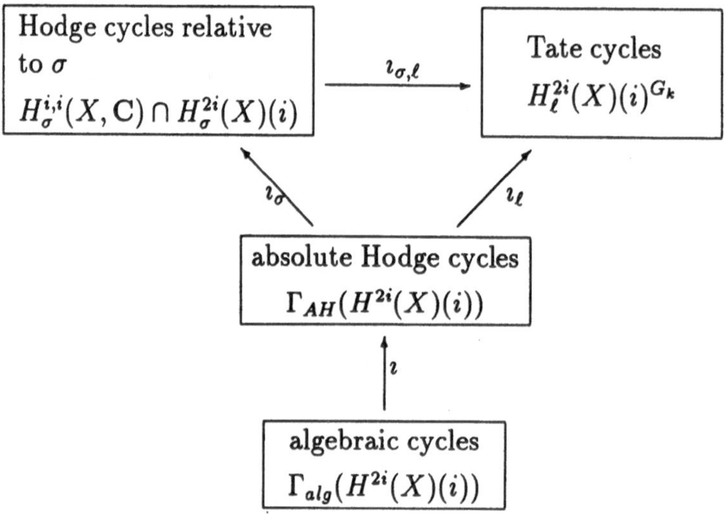

The maps $\imath$, $\imath_\sigma$ and $\imath_\ell$ are inclusions.

Conjecture 8.9.1 (Jannsen) $\gamma_X^i \otimes 1$ *is surjective.*

This conjecture is implied by either the Hodge or the Tate Conjecture. In conjunction, they lead to:

(*i*) If $k = \bar{k}$, Deligne's hope that $\imath_\sigma$ be surjective, in other words, that every Hodge cycle be an absolute Hodge cycle.

(*ii*) If k is finitely generated over its prime field, $\imath_\ell$ is surjective.

(*iii*) If k is finitely generated and big enough, $\imath_{\sigma,\ell} \otimes \mathbf{Q}_\ell$ is an isomorphism.

(*iv*) If k is finitely generated and big enough,

$$\dim_{\mathbf{Q}} H_\sigma^{i,i}(X) \cap H_\sigma^{2i}(X)(i) = \dim_{\mathbf{Q}_\ell} H_\ell^{2i}(X)(i)^{G_k}, \ \forall\sigma, \ \forall\ell.$$

Corollary 8.9.1

(a) (i) and (iii) would give: Tate $\Longrightarrow$ Hodge.

(b) (ii) and (iii) would give: Hodge $\Longrightarrow$ Tate.

Chapter 9

Mixed realizations, mixed motives and Hodge and Tate conjectures for singular varieties

This chapter is a continuation of the previous one. Jannsen's formalism of mixed realizations and mixed motives is discussed. These are defined as extensions of their pure analogues and the corresponding categories should be tannakian. Deligne has suggested a somewhat different definition of mixed motives, but in both Jannsen's and his conception the fundamental notion has become the realization.

Also, a Poincaré duality theory taking its values in the $\mathbf{Z}$*-linear tensor category of integral mixed realizations is constructed. The homological part of this duality theory is then used to state Hodge and Tate conjectures for arbitrary varieties. Again, for varieties over number fields, both conjectures are different aspects of the same underlying motivic formalism.*

A homological regulator for singular varieties is suggested to play the role of the usual regulator. This is in line with the general Riemann-Roch Theorem.

9.1 Tate Modules

As a first step we introduce the category $\mathcal{H}od_{\mathbf{Q}}$ of **Q-rational Hodge structures**. Objects of this category are finite dimensional $\mathbf{Q}$-vector spaces V with a real Hodge structure on $V \otimes \mathbf{R}$ whose weight filtration is defined over $\mathbf{Q}$. The morphisms are the obvious ones. This category $\mathcal{H}od_{\mathbf{Q}}$ is a (neutral) tannakian category. The Betti (singular) cohomology functor

$$H_\sigma : \mathcal{V}_k \longrightarrow \mathcal{H}od_{\mathbf{Q}}, \quad \sigma : k \longrightarrow \mathbf{C},$$

factorizes over $\mathcal{M}_k$ to give the realization functor, again denoted by H_σ,

$$H_\sigma : \mathcal{M}_k \longrightarrow \mathcal{H}od_{\mathbf{Q}}.$$

Deligne's 'espoir' is rephrased by conjecturing that the functor

$$H_\sigma : \mathcal{M}_k \longrightarrow \mathcal{H}od_{\mathbf{Q}}$$

is fully faithful. This is a direct consequence of the Hodge Conjecture translated to this language.

Deligne's result (Theorem 8.6.1) for abelian varieties can be reformulated as the fully faithfulness of H_σ on the tannakian subcategory $\mathcal{M}_k^{av}$ of $\mathcal{M}_k$ generated by Artin motives and motives of abelian varieties. Here the category of (E.)Artin motives $\mathcal{M}_k^0$ is the one generated by zero-dimensional varieties (i.e products of number fields) over k. By Grothendieck's formulation of Galois theory one has an equivalence of tensor categories

$$\mathcal{M}_k^0 \xrightarrow{\sim} \mathcal{R}ep_c(G_k, \mathbf{Q}) \xrightarrow{\sim} \mathcal{S}(\mathrm{Spec}(k)_{ét}, \mathbf{Q}),$$

where this last category is the category of sheaves of finite dimensional $\mathbf{Q}$-vector spaces on the étale site $\mathrm{Spec}(k)_{ét}$.

The category of effective motives of degree 1, $\mathcal{M}_k^{+1}$, is the Karoubian (pseudo-abelian) subcategory of $\mathcal{M}_k^+$ generated by the $h^1(X)$, where X is a geometrically connected curve over k. Manin [Man] proved that $\mathcal{M}_k^{+1}$ is equivalent to the category $\mathcal{I}sab_k$ of isogeny classes of abelian varieties over k, where the correspondence is given by associating to

$h^1(X)$, X a curve, its Jacobian Jac(X). For $k = \mathbf{C}$ one has a fully faithful functor given by Betti cohomology

$$H_B^1 : \mathcal{I}sab_{\mathbf{C}} \longrightarrow \mathcal{H}od_{\mathbf{Q}}\,.$$

Summarizing one can state:

Conjecture 9.1.1 (Deligne's 'espoir') *For any algebraically closed field k and embedding $\sigma : k \longrightarrow \mathbf{C}$, the tensor functor*

$$H_\sigma : \mathcal{M}_k \longrightarrow \mathcal{H}od_{\mathbf{Q}}$$

is fully faithful.

Theorem 9.1.1 (Deligne) *For any algebraically closed field k with embedding $\sigma : k \longrightarrow \mathbf{C}$, the tensor functor*

$$H_\sigma : \mathcal{M}_k^{av} \longrightarrow \mathcal{H}od_{\mathbf{Q}}$$

is fully faithful.

Examples of effective motives $h(X)$ belonging to $\mathcal{M}_k^{av}$ are provided by varieties X, where X is:

(*i*) a curve;

(*ii*) a unirational variety of dimension ≤ 3;

(*iii*) a Fermat hypersurface;

(*iv*) a K_3-surface.

For a field k of finite type over the prime field, e.g. a number field or a finite field, we denote by

$$\mathcal{T}ate(k) = \mathcal{R}ep_c(G_k, \mathbf{Q}_\ell)$$

the category of **Tate modules**, i.e. finite dimensional $\mathbf{Q}_\ell$-vector spaces with continuous G_k-action. This category is equivalent with the category of constructible $\mathbf{Q}_\ell$-sheaves on $\mathrm{Spec}(k)_{\acute{e}t}$:

$$\mathcal{T}ate(k) \xrightarrow{\sim} \mathcal{S}_c(\mathrm{Spec}(k)_{\acute{e}t}, \mathbf{Q}_\ell)\,.$$

It is (neutral) tannakian over $\mathbf{Q}_\ell$.

The category of **graded Tate modules** will be written $\mathcal{G}r\mathcal{T}ate(k)$. The ℓ-adic cohomology functor

$$H_\ell : \mathcal{V}_k \longrightarrow \mathcal{G}r\mathcal{T}ate(k)$$

induces a faithful $\mathbf{Q}$–linear tensor functor

$$H_\ell : \mathcal{M}_k \longrightarrow \mathcal{G}r\mathcal{T}ate(k)\,.$$

The following conjecture is due to Tate [Ta1]:

Conjecture 9.1.2 (Tate) *Let k be a field finitely generated over the prime field. Then, for $M, N \in \mathcal{O}b(\mathcal{M}_k)$, the inclusion*

$$\mathrm{Hom}_{\mathcal{M}_k}(M,N) \hookrightarrow \mathrm{Hom}_{\mathcal{G}r\mathcal{T}ate(k)}(H_\ell(M), H_\ell(N))$$

induces an isomorphism

$$\mathrm{Hom}_{\mathcal{M}_k}(M,N) \otimes \mathbf{Q}_\ell \xrightarrow{\sim} \mathrm{Hom}_{\mathcal{G}r\mathcal{T}ate(k)}(H_\ell(M), H_\ell(N))\,.$$

The following result is due to Tate (cf. [Ta2]) for finite fields and to Faltings (cf. [Fa]) for number fields.

Theorem 9.1.2 (Tate, Faltings) *Let k be a finite field or a number field and let X and Y be abelian varieties over k. Then the map*

$$\mathrm{Hom}_k(X,Y) \otimes \mathbf{Z}_\ell \longrightarrow \mathrm{Hom}_{G_k}(T_\ell(X), T_\ell(Y))$$

is bijective.

Here T_ℓ denotes the Tate module: for an abelian variety A, let $A[n]$ denote the kernel of multiplication by n on A. Then, by definition,

$$T_\ell(A) = \varprojlim_{\nu} A[\ell^\nu](\bar{k})\,.$$

Remark 9.1.1 Actually, the statement of the theorem can be replaced by an apparently weaker, but nonetheless equivalent statement, namely: the map

$$\mathrm{Hom}_k(X,Y) \otimes \mathbf{Q}_\ell \longrightarrow \mathrm{Hom}_{G_k}(V_\ell(X), V_\ell(Y))$$

is bijective, where $V_\ell = T_\ell \otimes_{\mathbf{Z}_\ell} \mathbf{Q}_\ell$. V_ℓ can be regarded as the dual of $H^1_{\text{ét}}(-, \mathbf{Q}_\ell)$, and one has an identification

$$\mathrm{Hom}_{G_k}(V_\ell(X), V_\ell(Y)) \xrightarrow{\sim} \mathrm{Hom}_{\mathcal{T}ate(k)}(H^1_\ell(Y), H^1_\ell(X))\,.$$

9.2 Mixed Realizations

From now on, unless otherwise stated, k will be a field of characteristic zero, embeddable into the complex numbers $\mathbf{C}$ by embeddings $\sigma : k \longrightarrow \mathbf{C}$, $\bar{\sigma} : \bar{k} \longrightarrow \mathbf{C}$ such that $\bar{\sigma}|k = \sigma$. As usual, $G_k = \mathrm{Gal}(\bar{k}/k)$ will denote the Galois group. To generalize the Hodge and Tate Conjectures to arbitrary varieties over k, the natural setting will be the homological one. Of course, for smooth projective varieties one may expect the statements to reduce to the aforementioned conjectures in terms of cohomology taking values in a suitable tensor category. Jannsen's approach is to construct a Poincaré duality theory with values in a $\mathbf{Z}$–linear tensor category with weights, the so-called category of **integral mixed realizations**, written as $\mathcal{IMR}_k$. The construction takes several steps.

The first step consists of the construction of the category $\mathcal{MR}_k$ of **mixed realizations** enlarging the category $\mathcal{R}$ of **realizations** of motives for absolute Hodge cycles, by introducing mixed structures, i.e. weight filtrations, on the various cohomology theories occurring in the realizations.

The objects of $\mathcal{MR}_k$ are families $H = (H_\sigma, H_{DR}, H_\ell; I_{\infty,\sigma}, I_{\ell,\bar{\sigma}})$, where ℓ runs over all primes and $\sigma, \bar{\sigma}$ run over all embeddings $k, \bar{k} \longrightarrow \mathbf{C}$. Each H_σ is a mixed $\mathbf{Q}$–Hodge structure, H_{DR} is a finite dimensional k-vector space with decreasing Hodge filtration and increasing weight filtration, and the H_ℓ are finite dimensional $\mathbf{Q}_\ell$-vector spaces with continuous G_k-action and increasing G_k-equivariant weight filtration. The $I_{\infty,\sigma}$ and $I_{\ell,\bar{\sigma}}$ are the obvious comparison morphisms respecting the various filtrations. Morphisms between two of these families are triples $(f_\sigma, f_{DR}, f_\ell)$ of morphisms between the respective components of the families, respecting the filtrations and corresponding under the comparison isomorphisms. The various weight filtrations combine to give a weight filtration $W_\bullet H$ on the objects $H \in \mathcal{O}b(\mathcal{MR}_k)$ and H is called **pure of weight** m if $W_m H = H$ and $W_{m-1}H = 0 = (0,0,0;0,0)$, the zero family. The category $\mathcal{R}_k$ is the full subcategory of $\mathcal{MR}_k$ whose objects are direct sums of pure realizations.

The objects of $\mathcal{MR}_k$ are successive extensions of pure realizations of the form $Gr^W_m H = W_m H / W_{m-1} H$. There is a natural tensor functor

$$\otimes : \mathcal{MR}_k \times \mathcal{MR}_k \longrightarrow \mathcal{MR}_k\,,$$

a unit element $1 = (\mathbf{Q}, k, \mathbf{Q}_\ell; id_{\infty,\sigma}, id_{\ell,\bar{\sigma}})$ pure of weight zero, an internal Hom written $\mathcal{H}om$, the notion of a dual $H^\vee = \mathcal{H}om(H, 1)$ and faithful $\mathbf{Q}$–linear tensor functors, called **fibre functors**

$$\omega_\sigma : \mathcal{MR}_k \longrightarrow \mathcal{V}ect_{\mathbf{Q}}\,,$$

where $\mathcal{V}ect_{\mathbf{Q}}$ is the category of finite dimensional $\mathbf{Q}$–vector spaces, given by $\omega_\sigma(H) = H_\sigma$.

Also, write $\Gamma_{AH}(H)$ for the absolute Hodge cycles of H, which in this formulation of an object $H \in \mathcal{O}b(\mathcal{MR}_k)$ will consist of triples

$$(z_\sigma, z_{DR}, z_\ell)_{\sigma,\ell} \in \prod_\sigma H_\sigma \times H_{DR} \times \prod_\ell H_\ell = \prod_\sigma H_\sigma \times H_{\mathbf{A}}\,,$$

such that under the comparison isomorphisms the z_σ map to the z_{DR} and the z_ℓ, respectively, and such that $z_{DR} \in F^0 H_{DR} \cap W_0 H_{DR}$.

One has an isomorphism

$$\mathrm{Hom}(H, H') \simeq \Gamma_{AH}(\mathcal{H}om(H, H'))\,, \text{ for } H, H' \in \mathcal{O}b(\mathcal{MR}_k)\,.$$

In short, $\mathcal{MR}_k$ is a (neutral) **tannakian category** and $\mathcal{R}_k$ is a (neutral) **tannakian subcategory**, closed under the formation of subquotients. There are natural base extension and restriction functors and there is a Tate object, pure of weight -2.

The next step is to define contravariant functors

$$H^n : \mathcal{V}^\circ_k \longrightarrow \mathcal{MR}_k\,,$$

where $\mathcal{V}^\circ_k \supset \mathcal{V}_k$ is the category of **smooth, quasi-projective** varieties over k. So, let $U \in \mathcal{O}b(\mathcal{V}^\circ_k)$, and define:

$$H^n(U) = (H^n_\sigma(U), H^n_{DR}(U), H^n_\ell(U); I_{\infty,\sigma}, I_{\ell,\sigma})$$

with

$$\begin{cases} H^n_\sigma(U) &=& H^n(\sigma U, \mathbf{Q})\,, \\ H^n_{DR}(U) &=& H^n_{DR}(U/k) = \mathbf{H}^n(U_{Zar}, \Omega^\bullet_{U/k}) \text{ and} \\ H^n_\ell(U) &=& H^n_{\acute{e}t}(U \otimes_k \bar{k}, \mathbf{Q}_\ell)\,. \end{cases}$$

One still has to give the filtrations and verify the compatibilities. The Hodge and weight filtrations on $H^n_{DR}(U)$ are the usual ones as in [De1]. They are defined over k.

For $H^n_\sigma(U)$ one uses the smooth compactification σX of σU and Leray's spectral sequence

$$E_2^{p,q} = H^p(\sigma X, \mathrm{R}^q j_* \mathbf{Q}) \Longrightarrow H^{p+q}(\sigma U, \mathbf{Q}) = E^{p+q}, \qquad (9.1)$$

for $j : U \longrightarrow X$ the embedding, to obtain the weight filtration $W_\bullet$ with $Gr^W_{n+k} E^n = E_\infty^{n-k,k}$. The Hodge filtration on $H^n_\sigma(U)$ is just the one on the analytic de Rham cohomology

$$H^n_{DR}(\sigma U^{an}) \xrightarrow{\sim} H^n_{DR}(\sigma U/\mathbf{C}) \xrightarrow{\sim} H^n_{DR}(U/k) \otimes_{k,\sigma} \mathbf{C},$$

which is actually algebraic.

The weight filtration on $H^n_\ell(U)$ is defined as for $H^n_\sigma(U)$ via the spectral sequence

$$E_2^{p,q} = H^p_{\acute{e}t}(X \otimes_k \bar{k}, \mathrm{R}^q \bar{j}_* \mathbf{Q}_\ell) \Longrightarrow H^{p+q}_{\acute{e}t}(U \otimes_k \bar{k}, \mathbf{Q}_\ell) = E^{p+q}. \qquad (9.2)$$

The comparison isomorphisms etc. are verified, (cf. [Ja2]). The functor

$$H : \mathcal{V}^\circ_k \longrightarrow \mathcal{MR}_k$$

is defined by $H(U) = \bigoplus_{n \geq 0} H^n(U)$.

Still following U. Jannsen we come to a possible candidate for the category of mixed motives over k.

Definition 9.2.1 (Jannsen) *The category $\mathcal{MM}_k$ of* mixed motives over k *(for absolute Hodge cycles) is the tannakian subcategory of the tannakian category of mixed realizations $\mathcal{MR}_k$, generated by the $H(U)$, $U \in \mathcal{O}b(\mathcal{V}^\circ_k)$.*

One shows that a mixed realization $H \in \mathcal{O}b(\mathcal{MR}_k)$ is a mixed motive if and only if H is a subquotient of some $H(U) \otimes H(V)^\vee = \mathcal{H}om(H(V), H(U))$, with $U, V \in \mathcal{O}b(\mathcal{V}^\circ_k)$.

Theorem 9.2.1 (Jannsen) *The following diagram of functors and fully faithful embeddings of categories commutes:*

$$\begin{array}{ccccc}
\mathcal{V}_k & \xrightarrow{\ h\ } & \mathcal{M}_k & \hookrightarrow & \mathcal{R}_k \\
\downarrow & & \downarrow & & \downarrow \\
\mathcal{V}^\circ_k & \xrightarrow{\ H\ } & \mathcal{MM}_k & \hookrightarrow & \mathcal{MR}_k \,.
\end{array}$$

Remark 9.2.1 The above construction can be generalized almost verbatim to the case of smooth separated and smooth proper varieties instead of smooth quasi-projective and smooth projective varieties, respectively.

Remark 9.2.2 The restriction to smooth, quasi-projective varieties in the definition of mixed motives may be too strong, and probably, singular varieties must be included in the formalism. Also, no mention was made of the crystalline realization which will have to be incorporated in a general theory of mixed motives. At the moment of writing these lines no completely satisfactory definition of a mixed motive is available, (cf. next remark).

Remark 9.2.3 In [De4] Deligne, from the outset, includes the crystalline aspect in what he calls a system of realizations. This is the analogue of a mixed realization with some additional compatibility conditions to take the crystalline cohomology into account. With some hesitation he then gives a provisional 'definition' of the category of (mixed) motives as the subcategory of the category of systems of realizations, generated by those systems that are of 'geometric origin'. This last notion, however, remains unexplained and actually, it is not clear what one should understand by 'geometric origin'. Analogously to the situation of absolute Hodge cohomology one should have the vanishing of the higher Ext's:

$$\mathrm{Ext}^i_{\mathcal{MM}}(\mathbf{Z}(0), -),\ i \geq 2,$$

and the ensuing short exact sequence. Anyhow, it is hoped that the realization functors define a fully faithful functor from the category of mixed motives to the one of systems of realizations. Thus, mixed motives form a tannakian category.

The final step in the construction of the category of **integral mixed realizations** $\mathcal{IMR}_k$ consists of the modification of $\mathcal{MR}_k$ by replacing the mixed **Q**–Hodge structures H_σ by mixed **Z**–Hodge structures, also denoted by H_σ, such that there exist morphisms

$$Gr^W_m H_{\sigma,\mathbf{Q}} \otimes Gr^W_m H_{\sigma,\mathbf{Q}} \longrightarrow 1_{\mathbf{Q}}(-m),\ \forall m \in \mathbf{Z},$$

inducing polarizations on the real Hodge structures (of pure weight m) $Gr_m^W H_{\sigma,\mathbf{Q}} \otimes \mathbf{R}$. The $\mathbf{Q}_\ell$-vector spaces with continuous G_k-action must be replaced by finitely generated $\mathbf{Z}_\ell$-modules M_ℓ with continuous G_k-action such that the $M_\ell \otimes \mathbf{Q}_\ell$ have G_k-equivariant weight filtrations, in a sense to be made more precise below. Also, the comparison isomorphisms must be integrally defined. Morphisms between two integral mixed realizations are defined in an evident fashion. The final category $\mathcal{IMR}_k$ becomes a $\mathbf{Z}$-linear tensor category with weights, cf. *infra.*

9.3 Weights

Our aim is to arrive at the construction of a Poincaré duality theory $(H^\bullet_{AH}(X,\star), H^{AH}_\bullet(X,\star))$ for arbitrary varieties X over k and taking its values in the category $\mathcal{IMR}_k$. For $U \in \mathcal{O}b(\mathcal{V}_k^\circ)$ one would like to have $H^n_{AH}(U,0) \otimes \mathbf{Q} = H^n(U)$ in the category $\mathcal{MR}_k$. To get such a duality theory, let $\mathcal{V}ar_k$ denote the category of varieties (i.e. separated, reduced algebraic schemes) over k, and define for $X \in \mathcal{O}b(\mathcal{V}ar_k)$, $Z \subset X$ closed,

$$H^i_{AH,Z}(X,j) = H^i_{AH,Z}(X)(j) \text{ and } H^{AH}_a(X,b) = H^{AH}_a(X)(-b)\,.$$

For the various realizations one takes:

(*i*) $(H^i_{AH}(X))_\sigma = H^i_{\sigma,Z}(\sigma X, \mathbf{Z})$ and $(H^{AH}_a(X))_\sigma = H^{BM}_a(\sigma X, \mathbf{Z})$, the Borel-Moore homology. These functors take their values in the Poincaré duality theory of mixed $\mathbf{Z}$–Hodge structures. For smooth X, they are related to Beilinson's absolute Hodge cohomology by the surjective maps

$$H^i_{\mathcal{H}}(\sigma X, \mathbf{Z}(j)) \longrightarrow \Gamma_{\mathcal{H}}((H^i_{AH}(X,j)_\sigma) = \mathrm{Hom}_{\mathcal{H}}(\mathbf{Z}, H^i(\sigma X, \mathbf{Z}(j)))\,.$$

(*ii*) $(H^i_{AH,Z}(X))_{DR} = H^i_{DR,Z}(X/k)$ and $(H^{AH}_a(X))_{DR} = H^{DR}_a(X/k)$ via simplicial resolutions, (cf. [Del]). One obtains a Poincaré duality theory with values in $\mathcal{V}ect_k$, the category of finite dimensional k-vector spaces.

(*iii*) $(H^i_{AH,Z}(X))_\ell = H^i_{\acute{e}t,\bar{Z}}(\bar{X}, \mathbf{Z}_\ell)$ and $(H^{AH}_a(X))_\ell = H^{\acute{e}t}_a(\bar{X}, \mathbf{Z}_\ell) =$ $= H^{-a}_{\acute{e}t}(\bar{X}, \mathrm{R}\bar{f}^!\mathbf{Z}_\ell)$, where $f : X \longrightarrow \mathrm{Spec}(k)$ is the structure map.. This gives a Poincaré duality theory with values in $\mathcal{R}ep_c(G_k, \mathbf{Z}_\ell)$, the category of finitely generated $\mathbf{Z}_\ell$-modules with continuous G_k-action. This category is an abelian, $\mathbf{Z}_\ell$-linear tensor category, equivalent to the category of constructible $\mathbf{Z}_\ell$-sheaves on Spec(k). Tensored with $\mathbf{Q}_\ell$ it is equivalent to the category of constructible $\mathbf{Q}_\ell$-sheaves on Spec(k) or to $\mathcal{R}ep_c(G_k, \mathbf{Q}_\ell)$, the category of finite dimensional $\mathbf{Q}_\ell$-vector spaces with continuous G_k-action.

Write $\mathcal{IMR}_k \otimes \mathbf{Q}$, $\mathbf{Q} = (\mathbf{Q}, k, \mathbf{Q}_\ell)$, for the $\mathbf{Q}$–linear category with objects $M \otimes \mathbf{Q}$ corresponding to $M \in \mathcal{O}b(\mathcal{IMR}_k)$, and morphisms defined by

$$\mathrm{Hom}_{\mathcal{IMR}_k \otimes \mathbf{Q}}(M \otimes \mathbf{Q}, M \otimes \mathbf{Q}) = \mathrm{Hom}_{\mathcal{IMR}_k}(M, N) \otimes \mathbf{Q}\,.$$

To say that a Poincaré duality theory, taking values in $\mathcal{IMR}_k \otimes \mathbf{Q}$, has **weights** will mean that $\mathcal{IMR}_k \otimes \mathbf{Q}$ has a weight filtration, and this means that there is a sequence W_m of exact subfunctors of $id_{\mathcal{IMR}_k \otimes \mathbf{Q}}$, $m \in \mathbf{Z}$, such that

(i) $W_\bullet$ is increasing, finite, exhaustive and separated for every $M \otimes \mathbf{Q} \in \mathcal{O}b(\mathcal{IMR}_k \otimes \mathbf{Q})$,

(ii) For $M \otimes \mathbf{Q}, N \otimes \mathbf{Q} \in \mathcal{O}b(\mathcal{IMR}_k \otimes \mathbf{Q})$ one has

$$W_m((M \otimes \mathbf{Q}) \otimes (N \otimes \mathbf{Q})) = \sum_{i+j=m} W_i(M \otimes \mathbf{Q}) \otimes W_j(N \otimes \mathbf{Q})\,.$$

By abuse of language one says that a Poincaré duality theory with values in $\mathcal{IMR}_k$ has **weights** if it has weights in $\mathcal{IMR}_k \otimes \mathbf{Q}$.

Remark 9.3.1 In equations (9.3) and (9.4) below some estimates for the weights occurring in the Poincaré duality theories discussed heretofore are mentioned. Postulating these in other situations, one may define Poincaré duality theories with values in an abelian $\mathbf{Z}_{(\ell)}$-linear tensor category $\mathcal{T}$, such that $\mathcal{T} \otimes \mathbf{Q}_{(\ell)}$ is rigid, with weights in a manner analogous to the case of $\mathcal{IMR}_k$ and $\mathcal{IMR}_k \otimes \mathbf{Q}$. In particular the various Poincaré duality theories which make up the whole $(H^\bullet_{AH}(X,\star), H^{AH}_\bullet(X,\star))$ with values in $\mathcal{IMR}_k$ will have weights.

For the mixed Hodge structures and the de Rham part of our theory the weight filtrations are well known (cf. [Del]), but for the ℓ-adic part a word of explanation may be appropriate. So, let $M_\ell \otimes \mathbf{Q}_\ell \in$

$\mathcal{O}b(\mathcal{R}ep_c(G_k, \mathbf{Z}_\ell)) \otimes \mathbf{Q}_\ell = \mathcal{O}b(\mathcal{R}ep_c(G_k, \mathbf{Q}_\ell))$. Then, at least for finitely generated k, $M_\ell \otimes \mathbf{Q}_\ell$ is said to have a weight filtration if there exists an integral domain R over $\mathbf{Z}[\frac{1}{\ell}]$ with field of fractions equal to k such that $M_\ell \otimes \mathbf{Q}_\ell$ extends to a constructible $\mathbf{Q}_\ell$-sheaf $\mathcal{F}$ on Spec(R), with an increasing, finite, exhaustive and separating filtration $W_m\mathcal{F}$ by constructible subsheaves such that the $Gr_m^W \mathcal{F}$ are pointwise pure of weight m. This means that there exists a connected, smooth scheme U over Spec$(\mathbf{Z}[\frac{1}{\ell}])$ and a sheaf $\mathcal{F}$ over U, with filtration $W_\bullet$, such that the eigenvalues of the geometric Frobenius F_x, for a closed point x of U, i.e. of the induced Frobenius automorphism F_x^* acting on the fibre $(Gr_m^W \mathcal{F})_x$, are algebraic numbers of absolute value $Nx^{m/2}$. For details we refer to [De3]. Such a weight filtration is unique if it exists. Writing $W\mathcal{R}ep_c(G_k, \mathbf{Q}_\ell)$ for the full subcategory of $\mathcal{R}ep_c(G_k, \mathbf{Q}_\ell)$ formed by the representations with a weight filtration, every morphism of $W\mathcal{R}ep_c(G_k, \mathbf{Q}_\ell)$ is strictly compatible with the weight filtration, and, actually, $W\mathcal{R}ep_c(G_k, \mathbf{Q}_\ell)$ becomes a rigid, abelian, $\mathbf{Q}_\ell$-linear tensor category with weights.

For the weights w occurring in the $H_\bullet^{AH}(X,\star) \otimes \mathbf{Q}$ listed in the construction of $(H_{AH}^\bullet(X,\star), H_\bullet^{AH}(X,\star))$ one proves that

$$\begin{cases} 2b - a \le w \le 2b & \text{for } a \le \dim(X)\,, \\ 2b - a \le w \le 2b - 2(a - \dim(X)) & \text{for } a \ge \dim(X)\,. \end{cases} \tag{9.3}$$

and for proper X/k the weights w occurring in the $H_{AH}^\bullet(X,\star)$ one has

$$\begin{cases} -2j \le w \le i - 2j & \text{for } i \le \dim(X)\,, \\ 2(i - \dim(X)) - 2j \le w \le i - 2j & \text{for } i \ge \dim(X)\,. \end{cases} \tag{9.4}$$

Now, let $W\mathcal{R}ep_c(G_k, \mathbf{Z}_\ell)$ be the subcategory of $\mathcal{R}ep_c(G_k, \mathbf{Z}_\ell)$ consisting of those objects M_ℓ such that $M_\ell \otimes \mathbf{Q}_\ell$ has a weight filtration. Then this category is a rigid, abelian and $\mathbf{Z}_\ell$-linear tensor category and the functors $H_{AH}^\bullet(X,\star)_\ell$ and $H_\bullet^{AH}(X,\star)_\ell$ actually take their values in $W\mathcal{R}ep_c(G_k, \mathbf{Z}_\ell)$ and form a Poincaré duality theory with weights. The comparison isomorphisms and the (compatibilities of the) weight filtrations are defined via smooth hypercoverings. Thus the formalism of integral mixed realizations can be applied to the category $\mathcal{V}ar_k$, (cf. [Ja2]). Summarizing one can state:

Theorem 9.3.1 (Jannsen) *$(H^{\bullet}_{AH}(X,\star), H^{AH}_{\bullet}(X,\star)$ is a Poincaré duality theory with supports with values in the $\mathbf{Z}$-linear tensor category with weights $\mathcal{IMR}_k$, on the category $\mathcal{V}ar_k$ of varieties over the finitely generated field k of characteristic zero, that can be embedded into $\mathbf{C}$.*

9.4 Hodge and Tate Conjectures

Take $X \in \mathcal{O}b(\mathcal{V}ar_k)$ and let $Z_j(X) = \bigoplus_{x \in X_{(j)}} \mathbf{Z}$ be the group of cycles of dimension j. Such a cycle is a finite formal sum of irreducible closed subvarieties $Z \subset X$ of dimension j, i.e a finite formal sum of $\{\bar{x}\}$, $x \in X_{(j)}$. For a Poincaré duality theory taking values in the tensor category $\mathcal{T}$ one has a cycle map, induced by the fundamental class η_Z of $Z \in Z_j(X)$, which, by the principal triviality of the Poincaré duality theory, factorizes through rational equivalence to give a map

$$cl_j : CH_j(X) \longrightarrow \Gamma_{\mathcal{T}} H_{2j}(X,j)\,,$$

where the class of Z in $CH_j(X)$ is mapped to the image of η_Z under

$$\Gamma_{\mathcal{T}} H_{2j}(Z,j) \longrightarrow \Gamma_{\mathcal{T}} H_{2j}(X,j)\,.$$

Example 9.4.1 $(H^{\bullet}_{AH}(X,\star), H^{AH}_{\bullet}(X,\star))$, and $\mathcal{T} = \mathcal{IMR}_k$ or $\mathcal{MR}_k$.

For $(H^{\bullet}_{AH}(X,\star), H^{AH}_{\bullet}(X,\star))$ write $\Gamma_{AH} = (\Gamma_\sigma, \Gamma_{DR}, \Gamma_\ell)$, where Γ_σ, Γ_{DR} and Γ_ℓ are the obvious restrictions to the various subcategories involved. The following conjecture is due to Jannsen and generalizes the Hodge and Tate conjectures to the case of singular varieties.

Conjecture 9.4.1 (Jannsen) *For any $X \in \mathcal{O}b(\mathcal{V}ar_k)$,*

(i) if $k = \mathbf{C}$, then for every $j \geq 0$, the map

$$cl_j \otimes \mathbf{Q} : CH_j(X) \otimes \mathbf{Q} \longrightarrow \Gamma_\sigma(H^{AH}_{2j}(X,j) \otimes \mathbf{Q}) = \Gamma_{\mathcal{H}}(H_{2j}(X,\mathbf{Q}(j)) =$$

$$= (2\pi i)^{-j} W_{-2j} H^{BM}_{2j}(X,\mathbf{Q}) \cap F^{-j} H^{BM}_{2j}(X,\mathbf{C})$$

is surjective.

(ii) if k is finitely generated, then for every $j \geq 0$, the map

$$cl_j \otimes \mathbf{Q}_\ell : CH_j(X) \otimes \mathbf{Q}_\ell \longrightarrow \Gamma_\ell(H^{AH}_{2j}(X,j) \otimes \mathbf{Q}_\ell)) = H^{\acute{e}t}_{2j}(\bar{X},\mathbf{Q}_\ell(j))^{G_k}$$

is surjective.

Remark 9.4.1 In case (ii) k may be of characteristic $p > 0$, and $p \neq \ell$.

Using Chow's lemma and resolution of singularities, Jannsen proves the following remarkable theorem:

Theorem 9.4.1 (Jannsen) (i) *Conjecture* (i) *above is true if and only if it is true for smooth, projective varieties over* $\mathbf{C}$.

(ii) *If k is finitely generated of characteristic zero, Conjecture* (ii) *above is true if and only if it is true for smooth, projective varieties over* k.

Remark 9.4.2 In the case of $\mathrm{char}(k) > 0$, Conjecture (ii) is true if it is true for a so-called good proper cover of X. By this is meant a smooth and proper X'' fitting in a diagram of the form

$$\begin{array}{ccc} X' & \stackrel{\alpha}{\hookrightarrow} & X'' \\ \downarrow{\scriptstyle \pi} & & \\ X & & \end{array}$$

where π is proper and surjective and α is an open immersion into the smooth and proper variety X''. For X'' to be a **good proper cover** (for the homology theory H_*) one demands that $W_0H_{2j}(X,j)$ be a direct factor of $H_{2j}(X'',j)$, via $\pi_*\circ\alpha^*$. In particular, X'' is a good proper cover if $H_{2j}(X'',j)$ is a semi-simple object of the tensor category $\mathcal{IMR}_k$ or $\mathcal{MR}_k$.

One can also generalize Tate's conjecture on the order of the pole of the L-function of a variety, i.e. part (iii) of Beilinson's Conjecture II (cf. Conjecture 6.1.2) , to the case of arbitrary varieties $X \in \mathcal{O}b(\mathcal{V}ar_k)$ for finitely generated k. To this end one has to adapt the definition of $L(X,s)$ for singular varieties X over k. This goes as follows:

First, fix an integer i, $0 \leq i \leq \dim(X)$.

Second, one extends $X \longrightarrow \mathrm{Spec}(k)$ to a morphism

$$f : \mathcal{X} \longrightarrow U_0\,,$$

where U_0 is a smooth scheme over $\mathbf{Z}[\frac{1}{\ell}]$ if $\mathrm{char}(k) = 0$ or over $\mathbf{F}_p$ if $\mathrm{char}(k) = p > 0$, with generic point $\eta = \mathrm{Spec}(k)$. Then for a suitable open subscheme $U \subset U_0$ one can extend the sheaf $H^{2i}_{ét,c} = H^{2i}_{ét,c}(X \otimes_k \bar{k}, \mathbf{Q}_\ell)$ on Spec(k), where the subscript c means compact support, to a mixed sheaf $\mathcal{F}$ of weights $\leq 2i$ over U. In fact one can take $\mathcal{F} = \mathrm{R}^{2i} f_! \mathbf{Z}_\ell | U$. For a point y in U, write Ny for the cardinality of the finite residue field $\kappa(y)$ of y, and also write $F_y \in \mathrm{Gal}(\overline{\kappa(y)}/\kappa(y))$ for the geometric Frobenius. These F_y act on the fibre $\mathcal{F}_y$ of $\mathcal{F}$ at y, i.e. the fibre at the geometric point

$$\mathrm{Spec}(\overline{\kappa(y)}) \longrightarrow U$$

over y together with the action of $\mathrm{Gal}(\overline{\kappa(y)}/\kappa(y))$. The L-function $L(X,s)$ is then defined by

Definition 9.4.1 *The L-function of X is given by:*

$$L(X,s) = \prod_{y \in |U|} \frac{1}{\det(1 - F_y \cdot (Ny)^{-s} | \mathcal{F}_y)}, \quad s \in \mathbf{C},$$

where $|U|$ means the set of closed points of U. $L(X,s)$ converges for

$$\Re(s) > i + \dim_a(k),$$

where $\dim_a(k)$ denotes the Kronecker dimension of k, i.e.

$$\dim_a(k) = \mathrm{tr.deg}(k) \; (+1 \text{ if } \mathrm{char}(k) = 0).$$

The case of a smooth, projective variety X over $\mathbf{Q}$ (or a number field, or, more generally, a finitely generated field) is recovered by taking the usual sheaf $H^{2i}_{ét}(\bar{X}, \mathbf{Q}_\ell)$ which, in this case, coincides with $H^{2i}_{ét,c}(\bar{X}, \mathbf{Q}_\ell)$, over Spec$(k)$, and then extending to a smooth, projective morphism

$$f : \mathcal{X} \longrightarrow U.$$

For the sheaf $\mathcal{F}$ one may take $\mathcal{F} = \mathrm{R}^{2i} f_* \mathbf{Q}_\ell$. This sheaf is smooth (lisse) by the smooth and proper base change theorem and one has a $\mathrm{Gal}(\overline{\kappa(y)}/\kappa(y))$-isomorphism

$$\mathcal{F}_y \xrightarrow{\sim} H^{2i}_{ét}(\mathcal{X} \times_U \overline{\kappa(y)}, \mathbf{Q}_\ell) = H^{2i}_{ét}(X_p \otimes \bar{\mathbf{F}}_p, \mathbf{Q}_\ell),$$

if $\kappa(y) \simeq \mathbf{F}_q$, $q = p^f$, $p \neq \ell$. Then $\mathcal{F}_y$ is of pure weight $2i$ by Deligne's result on the Weil Conjectures for smooth, projective varieties over finite fields. Using Grothendieck's specialization theorem one obtains the L-function $L(X,s)$ as defined in Chapter 3 (at least up to a finite number of factors, depending on the scheme U).

Tate's Conjectures for arbitrary varieties X in $\mathcal{V}ar_k$ become:

Conjecture 9.4.2 (Jannsen) *Let k be a finitely generated field and let $X \in \mathcal{O}b(\mathcal{V}ar_k)$. Then, for every $i \geq 0$,*

(i) $cl_i \otimes \mathbf{Q}_\ell : CH_i(X) \otimes \mathbf{Q}_\ell \longrightarrow \Gamma_\ell(H_{2i}^{AH}(X,i) \otimes \mathbf{Q}_\ell) = H_{2i}^{ét}(\bar{X}, \mathbf{Q}_\ell(i))^{G_k}$ $(\mathrm{char}(k) \neq \ell)$ *is surjective.*

(ii) $B_i(X) = \mathrm{Im}(cl_i)$ *is finitely generated and one has*

$$B_i(X) \otimes \mathbf{Q}_\ell \xrightarrow{\sim} H_{2i}^{ét}(\bar{X}, \mathbf{Q}_\ell(i))^{G_k}.$$

(iii) $\mathrm{ord}_{s=m} L(X,s) = -\mathrm{rank}(B_i(X))$, *where* $m = i + \dim_a(k)$.

Remark 9.4.3 Though the L-function $L(X,s)$ depends on the scheme U, the order of the pole at $s = i + \dim_a(k)$ does not. This follows from the fact that for two choices U and U' the quotient of the corresponding L-functions converges for $s \in \mathbf{C}$ such that $\Re(s) > \dim(U) - 1 + i$.

To end this section we mention two results, also due to Jannsen.

Theorem 9.4.2 (Jannsen) *Assume that X has a good proper cover X'' for $H_*^{ét}$ (e.g. if $\mathrm{char}(k) = 0$). Then:*

If Conjecture (ii) above is true for X'' and dimension i, and Conjecture (i) above holds for $X'' \times X''$ and dimension $d'' = \dim(X'')$, then

Conjecture (ii) is true for X and dimension i.

Theorem 9.4.3 (Jannsen) *Assume that X has a good proper cover X'' for $H_*^{ét}$. Then:*

If Conjecture (iii) is true for X'' and dimension i, and, if for no composition factor W of $H_c^{2i}(\bar{X}'', \mathbf{Q}_\ell)$ the associated L-function of W (defined by a formula similar to the one for $V = H_{ét}^{2i}(\bar{X}, \mathbf{Q}_\ell)$ above) has a zero at the point $s = i + \dim_a(k)$ (e.g. if $\mathrm{char}(k) = p > 0$), and if Conjecture (ii) above holds for X and X'' and dimension i, then

Conjecture (iii) above is true for X and dimension i.

9.5 The Homological Regulator

One can also try to extend part of Beilinson's conjectures to singular varieties, i.e. extend the assertions of the previous section to higher algebraic K-theory. This is made possible by Gillet's Riemann-Roch Theorem.

Let $X \in \mathcal{O}b(\mathcal{V}ar_k)$ admit an embedding into a smooth variety M. Then by the purity theorem (cf. Theorem 4.4.7) and the definition of the maps τ_i (cf. Theorem 4.7.1) there is a commutative diagram

$$\begin{array}{ccc} K_i'(X) & \xrightarrow{\tau_i} & \bigoplus_{j\geq 0} \Gamma_\alpha(H^{AH}_{i-2j}(X,j)) \\ \uparrow \wr & & \uparrow \mathrm{td}(M)\cap \\ K_i^X(M) & \xrightarrow{\mathrm{ch}_i^X} & \bigoplus_{k\geq 0} \Gamma_\alpha(H^{2k-i}_{AH,X}(M,k)) \,. \end{array}$$

Here $\alpha \in \{AH, \sigma, DR, \ell\}$. Actually one may take the general situation of any Poincaré duality theory with values in a suitable tensor category with weights, but then one should be more careful with the Γ-functors which need not be exact. One obtains morphisms of functors:

$$(ch_*, \tau_*) : (K_*(-), K_*'(-)) \longrightarrow (\Gamma_\alpha H^\bullet_{AH}(-,\square), \Gamma_\alpha H^{AH}_\bullet(-,\square)) \quad (9.5)$$

compatible with all functorialities of Poincaré duality theories, except that the fundamental class $[\mathcal{O}_X]$ of K-theory is not mapped to the fundamental class η_X of (co)homology theory. This is just the core of Riemann-Roch.

The **regulator maps**

$$r : H^i_{\mathcal{M},Z}(X, \mathbf{Q}(j)) \longrightarrow \Gamma_\alpha(H^i_{AH,Z}(X,j))$$

are induced by the ch_i. Their homological counterpart leads to morphisms

$$r' : H_a^{\mathcal{M}}(X, \mathbf{Q}(b)) \longrightarrow \Gamma_\alpha(H_a^{AH}(X,b)) \,.$$

These are induced by the τ_a. In this way (r, r') is a morphism of Poincaré duality theories.

The following conjecture was stated in [Ja2] :

Conjecture 9.5.1 (Jannsen) *Let $X \in \mathcal{O}b(\mathcal{V}ar_k)$. Then, for $i,j \in$ $\mathbf{Z}$:*

(*i*) *If k is a number field, the homological regulator map*

$$r' : H_i^{\mathcal{M}}(X, \mathbf{Q}(j)) \longrightarrow \Gamma_{AH}(H_i^{AH}(X,j) \otimes \mathbf{Q})$$

is surjective.

(*ii*) *If $k = \mathbf{C}$ and X can be defined over a number field, the homological regulator map*

$$r' : H_i^{\mathcal{M}}(X, \mathbf{Q}(j)) \longrightarrow \Gamma_{\sigma}(H_i^{AH}(X,j) \otimes \mathbf{Q}) = \Gamma_{\mathcal{H}}(H_i(X(\mathbf{C}), \mathbf{Q}(j)))$$

is surjective.

(*iii*) *If k is a global or a finite field, the homological regulator map*

$$r' \otimes \mathbf{Q}_\ell : H_i^{\mathcal{M}}(X, \mathbf{Q}(j)) \otimes \mathbf{Q}_\ell \to \Gamma_\ell(H_i^{AH}(X,j) \otimes \mathbf{Q}_\ell) = H_i^{\acute{e}t}(\bar{X}, \mathbf{Q}_\ell(j))^{G_k}$$

is surjective, provided $\mathrm{char}(k) \neq \ell$.

One has the partial results:

Theorem 9.5.1 (Jannsen) *Let $X \in \mathcal{O}b(\mathcal{V}ar_k)$ be of dimension d over k. Then:*

(*i*) *Part* (*ii*) *of the above conjecture holds if* $(i,j) = (0,0)$, $(2d-2, d-1)$, $(2d-1, d-1)$ *and* $(2d, d)$, *hence it is true for curves.*

(*ii*) *Part* (*iii*) *of the above conjecture holds if* $(i,j) = (0,0)$, $(2d-1, d-1)$ *and* $(2d, d)$, *hence it is true for curves.*

The proof of this theorem is based on results of Soulé on the homological version of Quillen's spectral sequence (Theorem 4.4.5).

Chapter 10

Examples and Results

In this final chapter we discuss in some detail several results related to the conjectures of the foregoing chapters. The first result concerns important work of B. Gross and D. Zagier on the Birch & Swinnerton-Dyer Conjectures. Next, an overview of Deligne's Conjecture on the L-function of an algebraic Hecke character is given. This conjecture is now a theorem, due to work of D. Blasius, G. Harder and N. Schappacher. The third and fourth sections treat Beilinson's results on regulators for Artin motives and modular curves, respectively. In this last situation a (possibly) general phenomenon occurs: only part of motivic cohomology is useful. This phenomenon was already encountered in the discussion of Ramakrishnan's result on the regulator map for Hilbert modular surfaces. In the last section a class of varieties is introduced for which the Hodge and Tate Conjectures are true. This result is due to U. Jannsen.

10.1 B & S-D revisited

As a first example we consider a consequence of the now famous theorem of B. Gross and D. Zagier for the Birch & Swinnerton-Dyer Conjectures. For simplicity we take as the base field the rational numbers $\mathbf{Q}$ though the theorem of Gross and Zagier can be formulated for arbitrary number fields.

The theorem works for modular curves, so let $E/\mathbf{Q}$ be a modular elliptic curve with conductor N. Recall that the truth of the Shimura-Taniyama-Weil Conjecture 2.6.2 would imply that every elliptic curve over $\mathbf{Q}$ is modular. Thus there is a (non-constant) morphism

$$\gamma : X_0(N) \longrightarrow E$$

defined over $\mathbf{Q}$, or, by functoriality, E is a direct factor of the Jacobian $J_0(N)$ of the modular curve $X_0(N)$. The L-function $L(E,s)$ is the L-function of a modular form f of weight 2 on $\Gamma_0(N)$, the congruence subgroup of $SL_2(\mathbf{Z})$ of level N, i.e. the subgroup of $SL_2(\mathbf{Z})$ given by

$$\Gamma_0(N) = \left\{ \begin{pmatrix} a & b \\ c & d \end{pmatrix} \in SL_2(\mathbf{Z}) \mid c \equiv 0 \pmod{N} \right\}.$$

Similarly one defines the group $\Gamma_1(N) \subset \Gamma_0(N)$ by requiring also $a \equiv d \equiv 1 \pmod{N}$.

The functional equation for $L(E,s)$ is (cf. Conjecture 2.3.1)

$$\Lambda(E,s) = w\,\Lambda(E, 2-s)\,, \text{ with } w = \pm 1\,,$$

and the modular form f satisfies the equation

$$f(\frac{-1}{Nz}) = w\,N\,z^2\,f(z)\,,\ z \in \mathcal{H}\,, \text{ the upper halfplane}.$$

The modular curve $X_0(N)$ is the natural compactification of the open modular curve $Y_0(N)$ that classifies pairs of elliptic curves (E', E'') together with a cyclic isogeny

$$\phi : E' \longrightarrow E''$$

of degree N. The set of complex points of $Y_0(N)$ can be identified with $\Gamma_0(N)\backslash\mathcal{H}$. Then, for the modular curve $X_0(N)$, this gives $X_0(N)(\mathbf{C}) \simeq \Gamma_0(N) \backslash \mathcal{H}^*$, where $\mathcal{H}^* = \mathcal{H} \cup \mathbf{P}^1(\mathbf{Q})$. The set

$$\Gamma_0(N) \backslash \mathbf{P}^1(\mathbf{Q}) = X_0(N)(\mathbf{C}) \backslash Y_0(N)(\mathbf{C})$$

consists of the cusps of $X_0(N)(\mathbf{C})$.

Let K be an imaginary quadratic number field with discriminant D prime to N, ring of integers $\mathcal{O}_K$ and class number $h(K) = h$.

Definition 10.1.1 *A* Heegner point $y \in Y_0(N)(\mathbf{C})$ *of discriminant* D *is a point* y *corresponding to a pair* (E', E'') *with* $\mathrm{End}(E') = \mathrm{End}(E'') = \mathcal{O}_K$.

Heegner points can exist only if D is a square modulo $4N$. There are $2^t h$ Heegner points rational over the Hilbert class field $H = K(j(E))$ of K. Here t denotes the number of distinct primes dividing N. In $\mathcal{H}$ a Heegner point is given by an equation $az^2 + bz + c = 0$, $a, b, c \in \mathbf{Z}$ such that $\gcd(a, b, c) = 1$, $D = b^2 - 4ac$, $a = Na'$, $\gcd(a', b, Nc) = 1$. A Heegner point y of discriminant D gives rise to a point $c = (y) - (\infty) \in J_0(N)(H)$, where ∞ is the cusp at infinity. One has an action of the Hecke algebra $\mathbf{T}$ and the Galois group $\mathrm{Gal}(H/K)$ on $J_0(N)(H)$ and a fundamental question is to determine the cyclic module spanned by c over the ring $\mathbf{T}[\mathrm{Gal}(H/K)]$, acting as endomorphisms of $J_0(N)(H)$.

Let $f = \sum_{n\geq 1} a_n \exp(2\pi i n z)$, $a_1 = 1$, be a normalized eigenform under the action of $\mathbf{T}$, and let χ be a complex character of $\mathrm{Gal}(H/K)$ which is canonically isomorphic to the class group $Cl(\mathcal{O}_K)$ of K, i.e. the group of fractional ideals of $\mathcal{O}_K$ modulo (non-zero) principal ideals. Its order is just the class number h. To an element $\sigma \in \mathrm{Gal}(H/K)$ corresponds an ideal class $\mathcal{A} \in Cl(K)$ and this class defines a theta-series

$$\vartheta_{\mathcal{A}}(z) = \sum_{n\geq 0} r_{\mathcal{A}}(n) \exp(2\pi i n z), \; r_{\mathcal{A}}(0) = \frac{1}{2u}, \; u = \#(\mathcal{O}_K^*/\{+1, -1\}),$$

and $r_{\mathcal{A}}(n)$, $n \geq 1$, is the number of integral ideals in the class of $\mathcal{A}$ with norm n. $\vartheta_{\mathcal{A}}(z)$ is a modular form of weight one on $\Gamma_1(D)$, with character

$$\varepsilon : (\mathbf{Z}/D\mathbf{Z})^* \longrightarrow \{+1, -1\}$$

associated to the quadratic extension $K/\mathbf{Q}$. Then one defines the L-function $L(f, \chi, s)$ as

$$L(f, \chi, s) = \sum_{\mathcal{A}} \chi(\mathcal{A})\, L_{\mathcal{A}}(f, s) \tag{10.1}$$

with

$$L_{\mathcal{A}}(f, s) = \sum_{\substack{n\geq 1\\ (n, DN)=1}} \varepsilon(n)\, n^{1-2s} \cdot \sum_{n\geq 1} a(n)\, r_{\mathcal{A}}(n)\, n^{-s}. \tag{10.2}$$

The functions $L_A(f,s)$ and $L(f,\chi,s)$ extend to entire functions, satisfy functional equations for $s \leftrightarrow 2-s$, and vanish at the point $s=1$. Define for the divisor class c of $(y)-(\infty)$ in $J_0(N)(H)$

$$c_\chi = \sum_\sigma \chi^{-1}(\sigma)\, c^\sigma ,$$

then $(c_\chi)^\tau = \chi(\tau)\, c_\chi$ for all $\tau \in \mathrm{Gal}(H/K)$, in other words, c_χ is an element in the χ-eigenspace V^χ of $J_0(N)(H) \otimes \mathbf{C}$. The endomorphisms in $\mathbf{T}$ are defined over $\mathbf{Q}$, so they commute with the action of $\mathrm{Gal}(H/K)$ and V^χ admits a spectral decomposition

$$V^\chi = \bigoplus_f V^{\chi,f} ,$$

with f an eigenform of $\mathbf{T}$ (as before). Let $c_{\chi,f}$ stand for the projection of c_χ onto $V^{\chi,f}$, and write

$$(f,g) = \int_{\Gamma_0(N)\backslash\mathcal{H}} f(z)\overline{g(z)}\, dx dy\,, \quad z = x+iy\,,$$

for Petersson's inner product.

The main result of Gross and Zagier [GZ] is the following theorem:

Theorem 10.1.1 (Gross-Zagier) *Let f be a normalized eigenform of the Hecke algebra $\mathbf{T}$ and let χ be a complex character of* $\mathrm{Gal}(H/K) \simeq Cl(\mathcal{O}_K)$. *Define c_χ and $c_{\chi,f}$ as above. Then:*

$$L'(f,\chi,1) = \frac{8\pi^2(f,f)}{hu^2\sqrt{|D|}} \cdot \hat{h}_H(c_{\chi,f})\,,$$

where $\hat{h}_H$ is the canonical Néron-Tate height on $J_0(N)$ over H. One has $\hat{h}_H = h\,\hat{h}_K = 2h\,\hat{h}_{\mathbf{Q}}$.

Writing $\omega_f = 2\pi i\, f(z)\, dz$ for the eigendifferential associated to f and also

$$|\omega_f|^2 = \int_{X_0(N)(\mathbf{C})} \omega_f \wedge i\overline{\omega_f}\,,$$

the formula above becomes

$$L'(f,\chi,1) = \frac{|\omega_f|^2}{u^2\sqrt{|D|}} \cdot \hat{h}_K(c_{\chi,f})\,. \tag{10.3}$$

Assume that $\gamma : X_0(N) \longrightarrow E$ maps the cusp at infinity to the origin of E. Also assume that $L(E,1) = 0$.

There is a Néron differential ω (unique up to sign) on E such that $\gamma^*(\omega) = c\omega_f$ for a constant $c > 0$ and a modular form (newform) f such that $L(E,s) = L(f,s)$. One has the relation between periods

$$|\omega|^2 = \iint_{E(\mathbf{C})} |\omega \wedge \bar{\omega}| = \frac{c^2|\omega_f|^2}{\deg(\gamma)}. \tag{10.4}$$

Let y be a Heegner point of discriminant D on $X_0(N)$. Then, by the addition on $E(H)$, the point

$$P_K = \sum_\sigma \gamma(y^\sigma) = \sum_\sigma (\gamma(y))^\sigma$$

belongs to $E(K)$. Up to sign it is independent of the choice of y. One has the formula $\hat{h}_K(P_K) = \hat{h}_K(c_{1,f}) \deg(\gamma)$, where $\chi = 1$ is the trivial character on $\mathrm{Gal}(H/K)$. We obtain the formula:

$$L'(E/K,1) = \frac{|\omega|^2\, \hat{h}_K(P_K)}{c^2u^2\sqrt{|D|}}. \tag{10.5}$$

Furthermore, under complex conjugation the point P_K goes to $w\, P_K$, where w is the sign in the functional equation of $\Lambda(E,s)$. Thus, when $w = -1$, which implies that $\mathrm{ord}_{s=1} L(E,s)$ is odd, one obtains a point $2P_K \in E(\mathbf{Q})$. When $w = +1$ one gets a point $2P_K$ on the 'D^{th} twist' E_D of E, given by the equation

$$E_D : D\, y^2 = 4x^3 + ax + b$$

if E is given by

$$E : y^2 = 4x^3 + ax + b\,, \ a,b \in \mathbf{Z}\,.$$

Gross and Zagier use this formula and a theorem of Waldspurger, which implies the existence of a D with the necessary properties and such that $L(E_D,1) \neq 0$, to prove the following theorem:

Theorem 10.1.2 (Gross-Zagier) *Let E be a modular elliptic curve over $\mathbf{Q}$ such that $L(E,1) = 0$. Then there is a rational point $P \in E(\mathbf{Q})$ such that*

$$L'(E,1) \approx_{\mathbf{Q}^\bullet} \Omega \, \hat{h}_{\mathbf{Q}}(P) = \Omega \, \langle P, P \rangle \, ,$$

where Ω is the real period. In particular:

(i) If $L'(E,1) \neq 0$, then $E(\mathbf{Q})$ contains elements of infinite order. In other words:

$$\mathrm{ord}_{s=1} L(E/\mathbf{Q}, s) = 1 \Longrightarrow \mathrm{rank}(E(\mathbf{Q})) \geq 1 \, .$$

(ii) If $L'(E,1) \neq 0$ and $\mathrm{rank}(E(\mathbf{Q})) = 1$, then

$$L'(E,1) \approx_{\mathbf{Q}^\bullet} \Omega \, R \, ,$$

where $R = R(E/\mathbf{Q})$ is the elliptic regulator.

Remark 10.1.1 This may be compared with a theorem of Coates and Wiles for the rank zero case. Their result says that for a CM-elliptic curve E, $\mathrm{ord}_{s=1} L(E/\mathbf{Q}, s) = 0$ implies that $\mathrm{rank}(E(\mathbf{Q})) = 0$. Thus the theorem of Gross and Zagier is the first proven evidence for the Birch & Swinnerton-Dyer Conjectures in case the Mordell-Weil group is infinite. Besides, it is not restricted to the subset of CM-curves, which is relatively small.

Remark 10.1.2 Another result supporting the Birch & Swinnerton-Dyer Conjectures is due to Greenberg. It says that for a CM-elliptic curve $E/\mathbf{Q}$ with odd $\mathrm{ord}_{s=1} L(E/\mathbf{Q}, s)$ either $\mathrm{rank}(E(\mathbf{Q})) \geq 1$ or $Ш(E)$ is infinite.

Remark 10.1.3 As already mentioned in the discussion on the Gauß Class Number Problem (cf. Conjecture 1.1.2), the theorem of Gross and Zagier gives an additional result leading to the solution of Gauß's Class Number Problem. More exactly, D. Goldfeld proved that the class number problem could be solved provided he had an example of an elliptic curve E over $\mathbf{Q}$ with $\mathrm{ord}_{s=1} L(E/\mathbf{Q}, s) = \mathrm{rank}(E(\mathbf{Q})) = 3$. The main theorem of Gross and Zagier provides such an example, namely:

$$E: \; -139y^2 = x^3 + 10x^2 - 20x + 8 \, .$$

10.2 Deligne's Conjecture

The second conjecture where a positive result has been established, is Deligne's conjecture on the critical values of Hecke L-functions of totally imaginary number fields. This result is due to D. Blasius [Bla] and to G. Harder and N. Schappacher [HS]. For totally real number fields the conjecture follows from results of C.L. Siegel, and for number fields that are neither totally real nor totally imaginary there are no critical values. The conjecture was formulated in [De2] in terms of motives (for absolute Hodge cycles) associated to algebraic Hecke characters Ψ of a totally imaginary number field k taking values in a totally imaginary number field E.

A continuous homomorphism $\Psi : \mathbf{A}_{k,f}^* \longrightarrow E^*$ is called an **algebraic Hecke character of (infinity) type** β, with $\beta : k^* \longrightarrow E^*$ an algebraic homomorphism, i.e. β is a homomorphism induced by a rational character $\beta : R_{k/\mathbf{Q}}(\mathbf{G}_m) \longrightarrow R_{E/\mathbf{Q}}(\mathbf{G}_m)$, if the restriction of Ψ to $k_f^* \hookrightarrow \mathbf{A}_{k,f}^*$ coincides with β, in other words, for $x \in k^*$, $\Psi(x_f) = \beta(x)$. Here the subscript f means that the components at infinity are (changed to) 1.

For any complex embedding $\tau \in \mathrm{Hom}(E, \mathbf{C})$ one has

$$\tau_\circ\beta(x) = \prod_\sigma \sigma(x)^{n(\sigma,\tau)},$$

where σ runs over the complex embeddings $\mathrm{Hom}(k, \mathbf{C})$ and the $n(\sigma, \tau)$ are integers, invariant under the action of $\mathrm{Gal}(\bar{\mathbf{Q}}/\mathbf{Q})$ on

$$\mathrm{Hom}(k, \bar{\mathbf{Q}}) \times \mathrm{Hom}(E, \bar{\mathbf{Q}}) = \mathrm{Hom}(k, \mathbf{C}) \times \mathrm{Hom}(E, \mathbf{C}).$$

The **weight** of Ψ, $w = w(\Psi)$, is the integer $n(\sigma, \tau) + n(c\sigma, \tau)$, where c is complex conjugation on $\bar{\mathbf{Q}}$. This does not depend on σ, τ.

Writing $\mathbf{N}$ for the absolute norm, one has the identity $\Psi\bar{\Psi} = \mathbf{N}^w$.

Any $\tau \in \mathrm{Hom}(E, \mathbf{C})$ defines a complex valued **Größencharacter**

$$\tau_\circ\Psi : \mathbf{A}_{k,f}^* \longrightarrow \mathbf{C}^*,$$

which, after embedding $\mathbf{A}_{k,f}^* \hookrightarrow \mathbf{A}_k^*$, leads to an extension, also denoted $\tau_\circ\Psi$, defined on the idèle class group $\mathbf{A}_k^*/k^*$ of k. This leads to an array of L-functions, called **Hecke L-functions**, $L(\Psi, s) = (L(\tau_\circ\Psi, s))_\tau$, where

$$L(\tau_\circ\Psi, s) = \prod_\wp \left(1 - \frac{(\tau_\circ\Psi)(\pi_\wp)}{N\wp^s}\right)^{-1}, \quad \Re(s) > \frac{w}{2} + 1. \tag{10.6}$$

In this Euler product expansion $\wp$ runs over all prime ideals $\wp$ of k for which $\Psi(\pi_\wp)$ does not depend on the choice of the uniformizing parameter $\pi_\wp$ of $k_\wp$. This condition can be translated in terms of a conductor of (a character associated to) Ψ and then the Euler product is taken over the prime ideals of k, i.e. of the ring of integers $\mathcal{O}_k$ of k, that do not divide this conductor.

$L(\Psi, s)$ takes values in $E\otimes_{\mathbf{Q}}\mathbf{C}$. Hecke L-functions admit an analytic continuation to the entire complex plane and they satisfy a functional equation.

Assume $w(\Psi) = -1$ and $n(\sigma,\tau) \in \{-1,0\}$, $\sigma \in \mathrm{Hom}(k,\mathbf{C})$ and $\tau \in \mathrm{Hom}(E,\mathbf{C})$. Without loss of generality one may assume that E is a CM-field, i.e. quadratic over a totally real subfield, generated by the Ψ-values on the finite idèles of k. Then, by a theorem of W. Casselman, there is an abelian variety A defined over k such that:

(*i*) $2\dim(A) = [E : \mathbf{Q}]$;

(*ii*) There is an isomorphism $E \xrightarrow{\sim} \mathbf{Q} \otimes_{\mathbf{Z}} \mathrm{End}_k(A)$, and

(*iii*) $M(\Psi) = h_1(A) = h^1(A)(1)$ is a motive of weight -1 for Ψ, i.e. $M = M(\Psi)$ is defined over k, has E-action on all its realizations in the various cohomology theories and such that, for all primes ℓ, $H_\ell(M)$ is an $E \otimes \mathbf{Q}_\ell$-module of rank 1 with $\mathrm{Gal}(\bar{\mathbf{Q}}/k)$ acting via Ψ.

The usual compatibilities under the various comparison isomorphisms should hold, in particular, E embeds into $\mathrm{End}(M)$, and the condition on the rank implies that $H_\sigma(M)$ is a one-dimensional E-vector space. The $M(\Psi)$ is uniquely determined up to isomorphism.

The case of arbitrary weights can be treated by taking tensor products of motives of the form $h_1(A)$ or $h^1(A)$.

Denote by $\mathcal{CM}_k$ the tannakian subcategory of $\mathcal{M}_k^{av}$ generated by the Artin motives over k and by motives of the form $h^1(A)$ for abelian varieties A over k with complex multiplication, in the sense that $\mathrm{End}_{\bar{k}}(A)$ contains a number field E of degree $[E : \mathbf{Q}] = 2\dim(A)$.

The category $\mathcal{CM}_k(E)$ will be the category of motives in $\mathcal{CM}_k$ that are equipped with an E-action, $E \hookrightarrow \mathrm{End}(M)$. For details we refer to [Sch]. One concludes:

Theorem 10.2.1 *For any algebraic Hecke character Ψ of the number field k with values in E, there exists a unique (up to isomorphism)* *motive* $M = M(\Psi) \in Ob(\mathcal{CM}_k(E))$ *which is a motive for* Ψ *in the sense explained above. If* Ψ' *is another such algebraic Hecke character, then:*

$$M(\Psi.\Psi') = M(\Psi) \otimes_E M(\Psi').$$

Also, $M(\mathbf{N}^n) = \mathbf{Q}(-n) \otimes_{\mathbf{Q}} k$.

For the Hodge decomposition of $H_\sigma(M) = H_\sigma(M(\Psi))$ one may remark that, for all $\sigma \in \mathrm{Hom}(k, \mathbf{C})$ and all $\tau \in \mathrm{Hom}(E, \mathbf{C})$, one has an inclusion

$$H_\sigma(M) \otimes_{E,\tau} \mathbf{C} \hookrightarrow H_\sigma^{n(\sigma,\tau),\, w-n(\sigma,\tau)},$$

where the $n(\sigma,\tau)$ are the integers determined by the infinity type of Ψ.

The comparison isomorphism

$$I : \bigoplus_\sigma H_\sigma(M) \otimes_{\mathbf{Q}} \mathbf{C} \xrightarrow{\sim} H_{DR}(M) \otimes_{\mathbf{Q}} \mathbf{C}$$

is an isomorphism of $k \otimes E \otimes \mathbf{C}$-modules of rank 1.

Let $\{e_\sigma\}$, σ running through $\mathrm{Hom}(k, \mathbf{C})$, be an E-basis of $H_\sigma(M)$, and write $e = (e_\sigma \otimes 1)_\sigma$. Then e is a generator of the free rank 1 $k \otimes E \otimes \mathbf{C}$-module $\bigoplus_\sigma H_\sigma(M) \otimes \mathbf{C}$.

Let ω be a $k \otimes_{\mathbf{Q}} E$-basis of $H_{DR}(M)$. Then one can define the **period**

$$p(\Psi) = p(M) = p(M(\Psi)) = \frac{\omega}{I(e)} \in (k \otimes E \otimes \mathbf{C})^* \simeq$$

$$\simeq (\mathbf{C}^*)^{\mathrm{Hom}(k,\mathbf{C}) \times \mathrm{Hom}(E,\mathbf{C})},$$

which, modulo $(E^*)^{\mathrm{Hom}(k,\mathbf{C})}.(k \otimes E)^*$ because of possible changes of the bases of $H_\sigma(M)$ and H_{DR}, respectively, depends only on Ψ. One has the identities $p(\Psi.\Psi') = p(\Psi).p(\Psi')$ and $p(\mathbf{N}^n) = (2\pi i)^{-n}$.

Write $p(\Psi; \sigma, \tau)$ or simply $p(\sigma,\tau)$ for the (σ,τ)-component of the image of $p(\Psi)$ under the isomorphism

$$(k \otimes E \otimes \mathbf{C})^* \simeq (\mathbf{C}^*)^{\mathrm{Hom}(k,\mathbf{C}) \times \mathrm{Hom}(E,\mathbf{C})}.$$

Thus one may write $p(\Psi)$ as an array

$$p(\Psi) = (p(\Psi; \sigma, \tau))_{\sigma,\tau} = (p(\sigma,\tau))_{\sigma,\tau} \tag{10.7}$$

with $p(\Psi; \sigma, \tau) = p(\sigma, \tau) \in \mathbf{C}^*$ for all $\sigma \in \mathrm{Hom}(k, \mathbf{C})$ and all $\tau \in \mathrm{Hom}(E, \mathbf{C})$.

Example 10.2.1 Let A be an abelian variety over k with complex multiplication by E as defined by an algebraic Hecke character Ψ of k with values in E. Thus $M(\Psi) = h_1(A)$ and, necessarily, $w(\Psi) = -1$. Assume $n(\sigma,\tau) = -1$, then there is a holomorphic 1-form $\omega_{\sigma,\tau} \in H^1_{DR}(A^\sigma/k^\sigma) \cap H^{1,0}$ with E-action given by $\varepsilon^*(\omega_{\sigma,\tau}) = \tau(\varepsilon)\omega_{\sigma,\tau}$, where $\varepsilon \in E \hookrightarrow \mathrm{End}_{k^\sigma}(A^\sigma)$. Take a rational cycle γ_σ such that $H_1^\sigma(A) = E.\gamma_\sigma$, then

$$p(\Psi;\sigma,\tau) = p(h_1(A);\sigma,\tau) = \int_{\gamma_\sigma} \omega_{\sigma,\tau}\,,$$

up to multiplication by $(E^*)^{\mathrm{Hom}(k,\mathbf{C})}\,(k\otimes E)^*$.

Remark 10.2.1 For the motives $M(\Psi)$ for Ψ of arbitrary weight, the calculation of $p(M(\Psi);\sigma,\tau)$ reduces to integration of holomorphic differentials on which E acts via τ or $c\tau$ and using the generalized Legendre relation for abelian varieties:

$$p(h_1(A);\sigma,\tau)\,p(h_1(A);\sigma,\tau c) \approx 2\pi i\,.$$

The L-function of $M(\Psi)$, $L(M(\Psi),s)$ is defined as the L-function of Ψ, i.e. $L(\Psi,s)$. From general results (cf. [De2]) it follows that it is equal to

$$L(\Psi,s) = L(R_{k/\mathbf{Q}}M(\Psi),s) = (L(\tau\circ\Psi,s))_\tau\,.$$

Deligne's conjecture for this situation becomes:

Conjecture 10.2.1 (Deligne) *For critical s one has*

$$L(\Psi,s) \approx_{E^*} c^+(\Psi,s)\,.$$

This means that for all $\tau : E \hookrightarrow \mathbf{C}$, there is an $x \in E^*$ such that

$$L(\tau\circ\Psi,s) = \tau(x)\,c^+(\Psi,\sigma;\tau)\,,$$

where $c^+(\Psi,\sigma;\tau)$ are the components of the Deligne period $c^+(\Psi,s)$.

To simplify notation we will assume that the critical value s is equal to 0, so we can write $c^+(\Psi)$, ... etc. for $c^+(\Psi,s)$, ... etc. Actually, by twisting this can always be obtained, so it is no loss of generality. With these notations Deligne's conjecture is sometimes written in the form:

$$\frac{L(\Psi,0)}{c^+(\Psi)} \in E \hookrightarrow E\otimes_{\mathbf{Q}}\mathbf{C}\,.$$

We are left with the description of the Deligne period

$$c^+(\Psi) = c^+(R_{k/\mathbf{Q}}M(\Psi)) \in (E \otimes \mathbf{C})^*/E^* .$$

It is obtained from the periods $p(\Psi; \sigma, \tau)$ by the formula:

$$c^+(\Psi) = \delta^+(\Psi). \left(\prod_{n(\sigma,\tau)<\frac{w}{2}} p(\Psi; \sigma, \tau) \right)_\tau . \tag{10.8}$$

For the definition of $\delta^+(\Psi) \in (E \otimes \mathbf{C})^*/E^*$ we refer to [Sch]. The final result is:

Theorem 10.2.2 (Blasius-Harder-Schappacher) *Let $s = 0$ be a critical value for Ψ, then*

$$L(\Psi, 0) \approx_{E^*} c^+(\Psi) .$$

10.3 Artin and Dirichlet Motives

Apart from the examples already mentioned in previous chapters, there are a few other cases where parts of Beilinson's conjectures are verified.

The first example to be discussed in this chapter concerns the zero-dimensional case, i.e. number fields, and more generally, Artin motives. In this case Beilinson's first conjecture comprises a conjecture of B. Gross on the values of the Artin L-function of a representation of the Galois group $\mathrm{Gal}(\bar{\mathbf{Q}}/\mathbf{Q})$.

Let k and E be (finite) number fields and denote by $\mathcal{M}_k^0(E)$ the category of Artin motives defined over k and with coefficients in E. If $M \in \mathcal{O}b(\mathcal{M}_k^0(E))$ is represented by the pair (X, p), where X is a zero-dimensional variety over k and

$$p : H^0(\bar{X}, \mathbf{Q}) \otimes E \longrightarrow H^0(\bar{X}, \mathbf{Q}) \otimes E$$

is a projector, i.e. a $\mathrm{Gal}(\bar{k}/k)$-homomorphism with $p^2 = p$, then one defines the motivic and Deligne cohomology of the motive M as follows:

Definition 10.3.1 *(i)* $H^1_{\mathcal{M}}(M, \mathbf{Q}(n)) = p^* H^1_{\mathcal{M}}(X, \mathbf{Q}(n)) \otimes_{\mathbf{Q}} E.$

(ii) $H^1_{\mathcal{D}}(M_{/\mathbf{R}}, \mathbf{R}(n)) = p^* H^1_{\mathcal{D}}(X_{/\mathbf{R}}, \mathbf{R}(n)) \otimes_{\mathbf{Q}} E.$

$H^1_{\mathcal{M}}(M, \mathbf{Q}(n))$ is an E-vector space and $H^1_{\mathcal{D}}(M_{/\mathbf{R}}, \mathbf{R}(n))$ is a free $\mathbf{R}\otimes E$-module. Similarly one defines $H^1_{\mathcal{M}}(X_{\mathbf{Z}}, \mathbf{Q}(n))$.

Dirichlet's Unit Theorem 1.3.1 and the Borel-Beilinson Regulator Theorem 5.2.2 carry over to the motivic formulation as follows:

Theorem 10.3.1 *Let $M = (X,p) \in Ob(\mathcal{M}^0_k(E))$. Then the regulator maps $r_{\mathcal{D}}$ for X induce isomorphisms of free $\mathbf{R}\otimes E$-modules:*

(i) $r_{\mathcal{D}} : (H^1_{\mathcal{M}}(M_{\mathbf{Z}}, \mathbf{Q}(1)) \oplus H(M)^{\Gamma}) \otimes \mathbf{R} \xrightarrow{\sim} H^1_{\mathcal{D}}(M_{/\mathbf{R}}, \mathbf{R}(1))$, *where* $H(M) = p^*(H^0(\bar{X}, \mathbf{Q}) \otimes E)$ *and* $\Gamma = \mathrm{Gal}(\bar{k}/k)$, $n = 1$.

(ii) $r_{\mathcal{D}} : H^1_{\mathcal{M}}(M_{\mathbf{Z}}, \mathbf{Q}(n)) \otimes \mathbf{R} \xrightarrow{\sim} H^1_{\mathcal{D}}(M_{/\mathbf{R}}, \mathbf{R}(n))$, $n > 1$.

The corresponding **regulator**, written $c_M(1-n)$, is defined as the determinant $\det(r_{\mathcal{D}})$ of the matrix associated to $r_{\mathcal{D}}$ after a choice of E-bases of the E-space $H^1_{\mathcal{M}}$ and the induced E-structure $H^1_{\mathcal{D}}$. Then $\det(r_{\mathcal{D}})$ is determined up to an element in E^*. Thus one has for the regulator:

$$c_M(1-n) \in (\mathbf{R}\otimes E)^*/E^*\,,\ n \geq 1\,. \tag{10.9}$$

For every $\tau \in \mathrm{Hom}(E, \mathbf{C})$, M defines a complex representation of the Galois group $\Gamma = \mathrm{Gal}(\bar{k}/k)$,

$$\rho_\tau : \Gamma \longrightarrow \mathrm{Aut}_{\mathbf{C}}(H(M) \otimes_{E,\tau} \mathbf{C})\,,$$

and therefore an Artin L-function $L(M_\tau, s)$. The L-series of the motive M is defined as the array

Definition 10.3.2 $L(M,s) = (L(M_\tau, s))_\tau$.

This L-function $L(M,s)$ is a meromorphic function with values in $\mathbf{C}\otimes_{\mathbf{Q}} E$. One can speak of the order of $L(M_\tau, s)$ at the point $s = m$. This does not depend on τ. Thus it makes sense to call this integer the **order** of $L(M,s)$ at the point $s = m$. The following result holds:

Theorem 10.3.2 *For every $n \geq 1$, one has*

$$\mathrm{ord}_{s=1-n} L(M,s) = \dim_E H^1_{\mathcal{M}}(M_{\mathbf{Z}}, \mathbf{Q}(n))\,.$$

Writing $M^\vee$ for the dual motive of M, i.e. the motive whose realizations are dual vector spaces or contragredient F_∞- or $\mathrm{Gal}(\bar{k}/k)$-modules to the vector spaces and F_∞- or $\mathrm{Gal}(\bar{k}/k)$-modules of the realizations of M, part (ii) of Beilinson's Conjecture I (cf. Conjecture 5.7.1) for the Artin motives with values in the number field E, reads as follows:

Conjecture 10.3.1 (Gross-Beilinson) *One has*

$$c_{M^\vee}(1-n) \approx_{E^*} L^*(M,s)_{s=1-n}\,.$$

Several special cases can be recognized:

Example 10.3.1 $M = (\mathrm{Spec}(k), id)$, the trivial motive $\underline{1}$, $E = \mathbf{Q}$ and $n = 1$, give

$$\zeta_k^*(0) \approx_{\mathbf{Q}^*} R,$$

where R is the classical regulator and the conjecture is just a weak form of the Class Number Formula. For $n > 1$ the conjecture is true by Borel's theorem 5.1.2.

Example 10.3.2 Another example arises when $1-n$ is critical. Then one recovers Deligne's conjecture for Artin motives which implies here that

$$L(M, 1-n) \in E^* \subset (\mathbf{C} \otimes E)^*,$$

where the embedding $E \hookrightarrow \mathbf{C} \otimes E$ is given by

$$a \mapsto (\ldots, \sigma a, \ldots)_{\sigma \in \mathrm{Hom}(E,\mathbf{C})}$$

after the identification $\mathbf{C} \otimes E \xrightarrow{\sim} \prod_\sigma \mathbf{C}$. The conjecture is true by results of Siegel.

Example 10.3.3 The general situation with $n = 1$ is known as Stark's Conjecture. It was proved by Tate [Ta6] in case the character of $H(M)$ has rational values. For an arbitrary Artin motive it remains a conjecture.

When $k = \mathbf{Q}$ and $X = \mathrm{Spec}(\mathbf{Q}(\zeta_N))$ one can associate to X a so-called Dirichlet motive, written $[\chi]$, and the corresponding Dirichlet L-function

$$L([\chi], s) = \left(\sum_{k=1}^{\infty} \frac{\chi_\tau(k)}{k^s} \right)_{\tau \in \mathrm{Hom}(E,\mathbf{C})}, \qquad (10.10)$$

where $\chi : (\mathbf{Z}/N\mathbf{Z})^* \longrightarrow E^*$ is an E-valued Dirichlet character (extended by 0 to $\mathbf{Z}/N\mathbf{Z}$). The Dirichlet motive $[\chi]$ is defined as
$[\chi] = (X, p_\chi)$, with $p_\chi = \frac{1}{\#(G)} \sum_{\sigma \in G} \chi^{-1}(\sigma)\,\sigma$, and $G = \mathrm{Gal}(\mathbf{Q}(\zeta_N)/\mathbf{Q})$.
The morphism p_χ is a projector in $\mathrm{End}_{E,\Gamma}(H^0(\bar{X}, \mathbf{Q}) \otimes E)$ and in the sum defining the p_χ, χ is considered as a character of G. By explicit calculations, making use of the polylogarithm function

$$\mathrm{Li}_s(z) = \sum_k \frac{z^k}{k^s},$$

Beilinson (cf. [Bel], [Ne2]) proved [1]:

Theorem 10.3.3 (Beilinson) *Let $[\chi]$ be a Dirichlet motive over* $\mathbf{Q}$ *with coefficients in E. Then, for every $n \geq 1$,*

$$c_{[\chi^\vee]}(1-n) \approx_{E^*} L^*([\chi], 1-n)\,.$$

10.4 Modular Curves

The next example concerns regulators of modular curves.

Let $\mathbf{A}_f$ denote the ring of finite $\mathbf{Q}$–adèles and write G_f for the group $GL_2(\mathbf{A}_f)$. For any compact open subgroup $K \subset G_f$ there is a modular curve M_K whose $\mathbf{C}$-points are given by

$$M_K(\mathbf{C}) = G_{\mathbf{Q}} \backslash \mathcal{H}^{\pm 1} \times G_f/K\,,$$

where $G_{\mathbf{Q}}$ denotes $GL_2(\mathbf{Q})$ and $\mathcal{H}^{\pm 1} = \mathbf{C} \backslash \mathbf{R}$. Write $\bar{M}_K$ for the smooth compactification of M_K, and denote by M_K^∞ the scheme of 'cusps' of the moduli problem under consideration, i.e. $M_K^\infty = \bar{M}_K \backslash M_K$ with reduced scheme structure. Then $\bar{M}_K$ is a smooth, projective (not necessarily geometrically connected) curve over $\mathbf{Q}$. Also, let $\bar{M}_{K,\mathbf{Z}}$ be a regular model of $\bar{M}_K$ over $\mathbf{Z}$. Such a model exists, (cf. [KM]).

Example 10.4.1 A special case is

$$K = K_N = \mathrm{Ker}(p_N : GL_2(\hat{\mathbf{Z}}) \longrightarrow GL_2(\mathbf{Z}/N\mathbf{Z}))\,,$$

[1] The cases $n = 1$ and $\chi = 1$ were known.

$N \geq 3$, giving the complex points

$$M_N(\mathbf{C}) = M_{K_N}(\mathbf{C}) = G_{\mathbf{Q}}\backslash\mathcal{H}^{\pm 1} \times G_f/K_N$$

of the moduli scheme M_N of elliptic curves with level N structure, i.e. an isomorphism $(\mathbf{Z}/N\mathbf{Z})^2 \xrightarrow{\sim} E[N]$, where $E[N]$ are the points of order N of E. The projective limit $M = \varprojlim M_N$, or more generally, $M = \varprojlim M_K$, where the limit is taken over the compact open subgroups $K \subset G_f$, is the moduli scheme for elliptic curves with so-called universal level structure $\hat{\mathbf{Z}}^2 \xrightarrow{\sim} \varprojlim E[N]$. Analogously, one defines $\bar{M} = \varprojlim \bar{M}_N$ (or $\varprojlim \bar{M}_K$).

Let $\pi_N : X_N \longrightarrow M_N$, $\pi_K : X_K \longrightarrow M_K$ and $\pi : X \longrightarrow M$, $X = \varprojlim X_N$, (or $X = \varprojlim X_K$) be the universal curves, and write $\pi_\ell : X^\ell \longrightarrow M$ for the ℓ-fold fibre product over M, $\ell \geq 0$, the Kuga-Sato schemes. Thus $X^0 = M$ and $X^1 = X$. M, $\bar{M}$ and X are schemes over the cyclotomic field $k = \mathbf{Q}(\zeta) = \cup_N \mathbf{Q}(\zeta_N)$ and there is a canonical left action of G_f on M and on $\bar{M}$ and of the semi-direct product $G_f \ltimes \mathbf{A}_f^2$ on X such that $G(\hat{\mathbf{Z}})$ and $\hat{\mathbf{Z}}^2$ act via the limits of actions on finite levels. The semi-direct product $G_f \ltimes \mathbf{A}_f^{2\ell}$ acts on X^ℓ such that for a compact open subgroup $K \subset G_f \ltimes \mathbf{A}_f^{2\ell}$, the quotient $K\backslash X^\ell$ is of finite type over $\mathbf{Q}$. Furthermore $X^\ell = \varprojlim K\backslash X^\ell$, where the limit is taken over all such K. Then for any contravariant functor H on the category of schemes of finite type over $\mathbf{Q}$, one defines

Definition 10.4.1 $H(X^\ell) = \varinjlim H(K\backslash X^\ell)$.

There is a natural action of $G_f \ltimes \mathbf{A}_f^{2\ell}$ on $H(X^\ell)$.

$\bar{M}$ defines a motive $h(\bar{M}) \in \mathcal{O}b(\mathcal{M}_k^{av}(\bar{\mathbf{Q}}))$ with $\bar{\mathbf{Q}}[G_f]$-decomposition

$$h(\bar{M}) = h^0(\bar{M}) \oplus h^1(\bar{M}) \oplus h^2(\bar{M}) \in \mathcal{O}b(\mathcal{M}_k(\bar{\mathbf{Q}}))$$

with $h^0(\bar{M}) = \bar{\mathbf{Q}}[\mathrm{Spec}(k)]$, $h^2(\bar{M}) = h^0(\bar{M})(-1)$ by Poincaré duality and $h^1(\bar{M}) = \bigoplus_\pi M_\pi \otimes \pi$, where the sum runs over all irreducible, parabolic $\bar{\mathbf{Q}}$–representations π of G_f of weight two, and where the M_π are motives such that

$$H^\bullet(M_\pi) = \mathrm{Hom}_{G_f}(\pi, H^\bullet(\bar{M} \otimes_{\mathbf{Q}} \bar{\mathbf{Q}}))$$

for the realizations $H^\bullet$. In particular, for the ℓ-adic realization one has a $\mathrm{Gal}(\bar{\mathbf{Q}}/\mathbf{Q})$-action on

$$H^1_{\acute{e}t}(M_\pi, \mathbf{Q}_\ell) = \mathrm{Hom}_{G_f}(\pi, H^1_{\acute{e}t}(\bar{M} \otimes_{\mathbf{Q}} \bar{\mathbf{Q}}, \bar{\mathbf{Q}}_\ell)),$$

and one can define the Artin L-function $L(M_\pi, s)$. Writing $m(\pi, K)$ for the dimension of the K-invariant subspace of π, one has a product expansion

$$L(\bar{M}_K, s) = \prod_\pi L(M_\pi, s)^{m(\pi,K)}. \tag{10.11}$$

Actually, the L-functions $L(M_\pi, s)$ are automorphic L-functions of the form $L(\check{\pi}, s)$, where $\check{\pi}$ is the representation contragredient to π. Such L-functions admit an analytic continuation to the entire complex plane and they satisfy a functional equation which implies that $L(\pi, s)$ have simple zeroes at the non-positive integers $m \leq 0$.

As usual, we write $L^*(\pi, s)_{s=m} = L^*(\pi, m) \in (\bar{\mathbf{Q}} \otimes \mathbf{R})^*$ for the first non-zero coefficient of the Taylor series expansion of $L(\pi, s)$ at $s = m$. So, here one has

$$L^*(\pi, s)_{s=m} = \frac{d}{ds}(L(\pi, s))_{s=m}, \quad m \leq 0.$$

The decomposition of $h^1(\bar{M})$ induces decompositions of the realizations $H^1_{DR}(\bar{M}) \otimes_{\mathbf{Q}} \bar{\mathbf{Q}} \supset \Omega^1(\bar{M}) \otimes_{\mathbf{Q}} \bar{\mathbf{Q}}$ and $H^1_B(\bar{M}, \mathbf{Q}(i)) \otimes_{\mathbf{Q}} \mathbf{Q}$ and the corresponding $\bar{\mathbf{Q}}$–spaces $\Omega^1(M_\pi)$ and $H^1_B(M_\pi, \mathbf{Q}(i))$ are one-dimensional by the 'multiplicity one theorem' of Jacquet-Langlands, (cf. [Gel]).

Returning to the schemes $\pi^\ell : X^\ell \longrightarrow M$, one has Gysin maps

$$\pi^\ell_{*,W} : H^a_{\mathcal{M}}(W \backslash X^\ell, \mathbf{Q}(b)) \longrightarrow H^{a-2\ell}_{\mathcal{M}}(M, \mathbf{Q}(b-\ell))$$

for any compact open subgroup W of $\mathbf{A}^{2\ell}_f$. For two such subgroups $W_1 \subset W_2 \subset \mathbf{A}^{2\ell}_f$ one has

$$\pi^\ell_{*,W_1} = \#(W_2/W_1)\, \pi^\ell_{*,W_2}$$

on

$$H^a_{\mathcal{M}}(W_2 \backslash X^\ell, \mathbf{Q}(b)) \subset H^a_{\mathcal{M}}(W_1 \backslash X^\ell, \mathbf{Q}(b)).$$

Beilinson introduces a one-dimensional G_f-module[2] $\mathcal{L}$, dual to the module of $\mathbf{Q}$–valued invariant measures on $\mathbf{A}_f^2$. Then the action of G_f on $\mathcal{L}$ is given by multiplication by $|\det|$. The module $\mathcal{L}$ defines a canonical $\mathcal{L}^{\otimes\ell}$–valued measure μ on $\mathbf{A}_f^{2\ell}$, and one obtains a G_f-map

$$\pi_*^\ell : H_{\mathcal{M}}^a(X^\ell, \mathbf{Q}(b)) \longrightarrow H_{\mathcal{M}}^{a-2\ell}(M, \mathbf{Q}(b-\ell)) \otimes \mathcal{L}^\ell ,$$

equal to $\mu(W).\pi_{*,W}^\ell$ on $H_{\mathcal{M}}^a(W\backslash X^\ell, \mathbf{Q}(b))$.

For all $\ell \geq 0$, the map $H_{\mathcal{M}}^2(\bar{M}, \mathbf{Q}(\ell+2)) \longrightarrow H_{\mathcal{M}}^2(M, \mathbf{Q}(\ell+2))$ is injective.

Definition 10.4.2 (Beilinson) *The parabolic second motivic cohomology space $H_{\mathcal{M}}^2(M, \mathbf{Q}(\ell+2))^{parab}$ of the modular curve M is*

$$\pi_*^\ell\Big(\{H_{\mathcal{M}}^{\ell+1}(X^\ell, \mathbf{Q}(\ell+1)), H_{\mathcal{M}}^{\ell+1}(X^\ell, \mathbf{Q}(\ell+1))\}\Big) \otimes \mathcal{L}^{-\ell} .$$

Here the curly braces $\{.,.\}$ denote pairing in higher K-theory, supposed to be well defined. The space $H_{\mathcal{M}}^2(M, \mathbf{Q}(\ell+2))^{parab}$ is a G_f-submodule of $H_{\mathcal{M}}^2(M, \mathbf{Q}(\ell+2))$. One also defines:

Definition 10.4.3 *For the compactified modular curve $\bar{M}$ one defines:*

$$H_{\mathcal{M}}^2(\bar{M}, \mathbf{Q}(\ell+2))^{parab} = H_{\mathcal{M}}^2(M, \mathbf{Q}(\ell+2))^{parab} \cap H_{\mathcal{M}}^2(\bar{M}, \mathbf{Q}(\ell+2)) .$$

Beilinson proves the following theorem, (cf. [Be2]):

Theorem 10.4.1 (Beilinson) *With the above notations one has:*
(i) *If $\ell = 0$, then $H_{\mathcal{M}}^2(\bar{M}, \mathbf{Q}(2))^{parab} \subset H_{\mathcal{M}}^2(\bar{M}, \mathbf{Q}(2))_{\mathbf{Z}}$.*
(ii) *If $\ell > 0$, then $H_{\mathcal{M}}^2(M, \mathbf{Q}(\ell+2))^{parab} = H_{\mathcal{M}}^2(\bar{M}, \mathbf{Q}(\ell+2))^{parab} \subset$*
$\subset H_{\mathcal{M}}^2(\bar{M}, \mathbf{Q}(\ell+2))_{\mathbf{Z}} = \mathrm{Im}(H_{\mathcal{M}}^2(\bar{M}_{\mathbf{Z}}, \mathbf{Q}(\ell+2)) \longrightarrow H_{\mathcal{M}}^2(\bar{M}, \mathbf{Q}(\ell+2)))$.

The regulator map $r_{\mathcal{D}}$ defines a G_f-invariant subspace

$$\mathcal{P}_\ell = r_{\mathcal{D}}(H_{\mathcal{M}}^2(\bar{M}, \mathbf{Q}(\ell+2))^{parab}) \otimes_{\mathbf{Q}} \bar{\mathbf{Q}} \subset H_{\mathcal{D}}^2(\bar{M}_{/\mathbf{R}}, \mathbf{R}(\ell+2)) \otimes_{\mathbf{Q}} \bar{\mathbf{Q}} =$$

$$= H_B^1(\bar{M}_{/\mathbf{R}}, \mathbf{R}(\ell+1)) \otimes_{\mathbf{Q}} \bar{\mathbf{Q}}. \qquad (10.12)$$

By the decomposition of $h^1(\bar{M})$ as above, this means that there is a $\bar{\mathbf{Q}}$–subspace $\mathcal{P}_{\ell,\pi} \subset H_B^1(M_\pi, \mathbf{R}(\ell+1))$ such that $\mathcal{P}_\ell = \bigoplus_\pi \mathcal{P}_{\ell,\pi} \otimes \pi$.

Beilinson's Main Theorem is:

---

[2] Beilinson uses the notation ν for $\mathcal{L}$.

Theorem 10.4.2 (Beilinson) *One has*

$$\mathcal{P}_{\ell,\pi} = L^*(\pi, s)_{s=-\ell}.H_B^1(M_\pi, \mathbf{Q}(\ell+1)).$$

It may be remarked that one can likewise give analogous statements for the modular curves $\bar{M}_N$ of finite level, or modular curves $\bar{M}_K$ for arbitrary compact open subgroups $K \subset G_f$. With this in mind, both theorems above together for $\bar{M}_K$ imply the following corollary:

Corollary 10.4.1 *For any compact open subgroup $K \subset G_f$ and any integer $\ell \geq 0$, there is a subspace $\mathcal{P}_{K,\ell} \subset H^2_{\mathcal{M}}(\bar{M}_K, \mathbf{Q}(\ell+2))$ such that:*

(i) $r_{\mathcal{D}}(\mathcal{P}_{K,\ell})$ *is a* $\mathbf{Q}$*-structure of* $H^2_{\mathcal{D}}(\bar{M}_{K/\mathbf{R}}, \mathbf{R}(\ell+2))$*;*

(ii) $\det(r_{\mathcal{D}}(\mathcal{P}_{K,\ell})) \approx_{\mathbf{Q}^*} L^*(\bar{M}_K, s)_{s=-\ell}$*;*

(iii) $\mathcal{P}_{K,\ell} \subset H^2_{\mathcal{M}}(\bar{M}_K, \mathbf{Q}(\ell+2))_{\mathbf{Z}}$.

Remark 10.4.1 For a detailed proof of the corollary in case $\ell = 0$ we refer to the contribution of N. Schappacher and A. Scholl in [RSS]. An important ingredient in this proof is a theorem due to Manin and Drinfeld, which says that the subgroup of $\mathrm{Pic}^0(\bar{M}_K \otimes \mathbf{C})$ consisting of divisor classes with support on the cusps of $\bar{M}_K \otimes \mathbf{C}$ is finite. In particular, the difference of any two cusps is torsion in the Jacobian of $\bar{M}_K$, and this makes possible the line of thought of Remark 5.7.3.

The idea is to evaluate the (projection of) the regulator $r_{\mathcal{D}}(u, v)$, $u, v \in \mathcal{O}^*(\bar{M}_K) \otimes \bar{\mathbf{Q}}$, on a 1-form $\omega \in \Omega^1(\bar{M}_K \otimes \bar{\mathbf{Q}})$ to obtain an integral

$$\frac{1}{2\pi i} \int_{M_K(\mathbf{C})} \log|u| \, \overline{d\log(v)} \wedge \omega .$$

The problem is to show that this integral is contained in

$$\Omega.L'(\check{\pi}, 0)\,\bar{\mathbf{Q}} \subset \bar{\mathbf{Q}} \otimes \mathbf{C},$$

where $\Omega \in (\bar{\mathbf{Q}} \otimes \mathbf{C})^*$ is the positive Deligne period of ω. The integral does not vanish for suitable $K \in G_f$, $u, v \in \mathcal{O}^*(\bar{M}_K) \otimes \bar{\mathbf{Q}}$ and $\omega \in V_\pi^K$, the $m(\pi, K)$-dimensional, K-invariant subspace of the decomposition

$$\Omega^1(\bar{M}_K) \otimes \bar{\mathbf{Q}} = \bigoplus_\pi V_\pi^K$$

into the sum of pairwise non-isomorphic modules for the Hecke algebra. The integral can be written as an integral of an Eisenstein series, an associated anti-holomorphic 1-form, with at most simple poles at the cusps, and the parabolic weight two form ω. It can be evaluated by 'Rankin's trick'. The result is that the integral is contained in the $\bar{\mathbf{Q}}$-subspace of $\bar{\mathbf{Q}} \otimes \mathbf{C}$ generated by elements of the form predicted by the functional equation, so using this functional equation the desired result follows.

The proof of the integrality statement (iii) is based on an analysis of the supersingular points of the reductions modulo p of the model $M_{N,\mathbf{Z}}$ over $\mathrm{Spec}(\mathbf{Z})$.

In the general situation ($\ell \geq 0$) modular forms of higher weights are used. Beilinson defines a space $\tilde{M}^{\infty}$ over the cusps M^{∞} and for any cohomology theory $\mathcal{C}$ a residue map

$$\mathrm{res}^{\ell}_{\mathcal{C}} : H^{\ell+1}_{\mathcal{C}}(X^{\ell}, A(\ell+1)) \longrightarrow H^{0}_{\mathcal{C}}(\tilde{M}^{\infty}, A),$$

with $A = \mathbf{Q}$ or $\mathbf{R}$. For $\mathcal{C}$ one will need motivic $\mathcal{M}$, absolute Hodge $\mathcal{H}$ or Betti B cohomology. One can also define a subspace $\mathcal{F}^{\ell}_{A} \subset H^{0}_{B}(\tilde{M}^{\infty}, A)$ which can be identified with the space of all locally constant A-valued functions on G_f with suitable transformation properties under a certain triangular subgroup related to M^{∞}. The image of res^{ℓ}_{B} is contained in $\mathcal{F}^{\ell}_{A}$.

Similarly one defines $\mathcal{F}^{\ell} \subset H^{0}_{\mathcal{M}}(\tilde{M}^{\infty}, \mathbf{Q}) = H^{0}(\tilde{M}^{\infty}, \mathbf{Q})$. Then $\mathcal{F}^{\ell} \subset \mathcal{F}^{\ell}_{\mathbf{Q}}$ is the subspace invariant under the right action of $\hat{\mathbf{Z}}^{*}$.

By means of Eisenstein series one constructs a map

$$E^{\ell} : \mathcal{F}^{\ell}_{\mathbf{R}} \longrightarrow H^{\ell+1}_{DR}(X^{\ell}_{/\mathbf{R}}) = H^{\ell+1}_{B}(X^{\ell}, \mathbf{C})$$

which commutes with the G_f-action and such that one has, for A equal to $\mathbf{Q}$ or $\mathbf{R}$, $E^{\ell}(\mathcal{F}^{\ell}_{A}) \subset H^{\ell+1}_{B}(X^{\ell}, A(\ell+1))$. More precisely

$$\mathcal{F}^{\ell}_{A} \underset{\mathrm{res}^{\ell}}{\overset{E^{\ell}}{\rightleftarrows}} H^{\ell+1}_{B}(X^{\ell}, A(\ell+1)) \cap F^{\ell+1} H^{\ell+1}_{DR}(X^{\ell}_{/\mathbf{R}})$$

are mutually inverse G_f-isomorphisms. A differential form representing the cohomology class $E^{\ell}(\phi)$ has logarithmic singularities at infinity.

In absolute Hodge cohomology Beilinson constructs a right inverse to $\mathrm{res}^{\ell}_{\mathcal{H}}$,

$$E^{\ell}_{\mathcal{H}} : \mathcal{F}^{\ell}_{\mathbf{R}} \longrightarrow H^{\ell+1}_{\mathcal{H}}(X^{\ell}, \mathbf{R}(\ell+1)),$$

also by means of Eisenstein series. This $E^{\ell}_{\mathcal{H}}$ is related to E^{ℓ} by the splitting morphism S of the canonical short exact sequence

$$0 \longrightarrow H^{\ell}_{B}(X^{\ell}, \mathbf{R}(\ell)) \longrightarrow H^{\ell+1}_{\mathcal{H}}(X^{\ell}, \mathbf{R}(\ell+1)) \overset{S}{\rightleftarrows} H^{\ell+1}_{B}(X^{\ell}, \mathbf{R}(\ell+1)) \cap$$
$$\cap\, H^{\ell+1}_{DR}(X^{\ell}_{/\mathbf{R}}) \longrightarrow 0.$$

by the formula: $E^{\ell}_{\mathcal{H}} = S \circ E^{\ell}$.

Denote by $\omega^{\ell} = (\omega^1)^{\otimes \ell} = \pi^{\ell}_{*}(\Omega^{\ell}_{X^{\ell}/M})$ the sheaf of weight ℓ modular forms on M and, for $\phi \in \mathcal{F}^{\ell}_{\mathbf{R}}$, by $\mathcal{E}^{\ell}(\phi)$ the section of ω^{ℓ} on M which is the $(\ell, 0)$-component of the form $E^{\ell}_{\mathcal{H}}(\phi)$ along the fibres of π. One has

$$\pi^{\ell}_{*}(E^{\ell}_{\mathcal{H}}(\phi_1) \cup E^{\ell}_{\mathcal{H}}(\phi_2)) = E^{\ell}(\phi_1).\bar{\mathcal{E}}^{\ell}(\phi_2) - (-1)^{\ell} E^{\ell}(\phi_2).\bar{\mathcal{E}}^{\ell}(\phi_1)$$

in $H^2_{\mathcal{H}}(M, \mathbf{R}(\ell+2)) = H^1_B(M, \mathbf{R}(\ell+1))$ and this form is the restriction to M of a closed current, also written $\pi^{\ell}_{*}(E^{\ell}_{\mathcal{H}}(\phi_1) \cup E^{\ell}_{\mathcal{H}}(\phi_2))$, on $\bar{M}$. Then one can show that the subspace

$$r_{\mathcal{D}}(H^2_{\mathcal{M}}(M, \mathbf{Q}(\ell+2))^{parab}) \subset H^1_B(\bar{M}, \mathbf{R}(\ell+1))$$

is generated by elements $\pi^{\ell}_{*}(E^{\ell}_{\mathcal{H}}(\phi_1) \cup E^{\ell}_{\mathcal{H}}(\phi_2))$, where the ϕ_i, $i = 1, 2$, run through

$$\mathrm{res}^{\ell}_{\mathcal{M}}(H^{\ell+1}_{\mathcal{M}}(X^{\ell}, \mathbf{Q}(\ell+1))) \subset \mathcal{F}^{\ell} \subset \mathcal{F}^{\ell}_{\mathbf{Q}}.$$

In fact, $\mathrm{res}^{\ell}_{\mathcal{M}}$ maps onto $\mathcal{F}^{\ell}$.

The evaluation of $\pi^{\ell}_{*}(E^{\ell}_{\mathcal{H}}(\phi_1) \cup E^{\ell}_{\mathcal{H}}(\phi_2))$ on a holomorphic 1-form ω on $\bar{M}$ is equivalent to the calculation of Petersson's scalar products

$$(\omega, E^{\ell}(\phi_1)\, \bar{\mathcal{E}}^{\ell}(\phi_2))$$

for the $\omega \in V_{\pi}$ occurring in the Hecke decomposition

$$\Omega^1(\bar{M}) \otimes \bar{\mathbf{Q}} = \bigoplus_{\pi} V_{\pi}.$$

Here again, Rankin's trick is used and a result analogous to the one mentioned in the case $\ell = 0$ is found.

Remark 10.4.2 The method for $\ell > 0$ can also be applied to the case $\ell = 0$, but to fill in the details, some minor changes are necessary and, as already mentioned, the Drinfeld-Manin Theorem plays a key role in this case.

10.5 Other Modular Examples

Apart from the example of the last section there are a few more examples where some of Beilinson's conjectures have been verified. They all show a common feature, namely the existence of a $\mathbf{Q}$–subspace of the motivic cohomology spaces $H^i_{\mathcal{M}}(X, \mathbf{Q}(j))$ giving rise, via the regulator map $r_{\mathcal{D}}$, to a $\mathbf{Q}$–structure on the corresponding Deligne-Beilinson cohomology $H^i_{\mathcal{D}}(X_{/\mathbf{R}}, \mathbf{R}(j))$ with volume $\det(r_{\mathcal{D}})$ equal to the first non-vanishing coefficient of the L-function of X at a suitable integer value of the argument, up to a non-zero rational factor.

In all these examples the variety X is a modular curve or a Shimura curve, a product of such curves, a Hilbert modular surface or a product of elliptic modular surfaces. In the cases of products of curves and of surfaces one obtains a proof for the Tate conjectures in the same spirit as the one sketched for Hilbert modular surfaces in Chapter 6. For more details we refer to [Ra3] and the literature cited there.

10.6 Linear Varieties

Our final example is concerned with the Hodge and Tate Conjectures as formulated in Conjecture 9.4.1 and Conjecture 9.4.2. Besides the well known examples where the conjectures are true there is a whole class of so-called **linear varieties** whose homologies carry mixed structures with very simple structures on the graded pure pieces. We follow [Ja2] and define:

Definition 10.6.1 *For a scheme S, a flat S-scheme X is called* 0-linear *if $X = \emptyset$ or X is isomorphic with the affine S-space $\mathbf{A}^N_S$ for some $N \geq 0$. For $n \geq 1$, X is called* n-linear *if there is a triple (U, Y, Z) of flat S-schemes such that $Z \subseteq Y$ is a closed S-immersion and $U \subseteq Y$ is the open complement, Z and one of (U, Y) is $(n-1)$-linear and X is the other member of (U, Y). X is called* linear *if X is n-linear for some $n \geq 0$. In particular, X is n-linear implies that X is $(n+1)$-linear.*

Example 10.6.1 X a flat S-scheme stratified by affine S-spaces, e.g. $X = \mathbf{P}^N_S$, or, if $S = \mathrm{Spec}(k)$ for some field k, then examples of k-linear varieties are furnished by complements in $\mathbf{P}^N_k$ (or $\mathbf{A}^N_k$) of a union of linear subspaces, or successive blow-ups of $\mathbf{P}_k^N$ in linear subspaces, Grassmannians, flag varieties or varieties stratified by such varieties.

Theorem 10.6.1 (Jannsen) *Let* $X \longrightarrow \mathrm{Spec}(k)$ *be a linear variety over the finitely generated field* k. *Then:*

(i)(a) If $\ell \neq char(k)$, *for all* $m, a \in \mathbf{Z}$,

$$Gr_m^W H_a^{\acute{e}t}(\bar{X}, \mathbf{Q}_\ell) = \begin{cases} 0 & , \quad m \text{ odd} \\ \bigoplus \mathbf{Q}_\ell(-\nu) & , \quad m = 2\nu \,,\ \nu \in \mathbf{Z}\,, \end{cases}$$

as G_k*-module.*

(b) If $char(k) = 0$, *for all* $m, a \in \mathbf{Z}$,

$$Gr_m^W H_a^{AH}(X) \otimes_{\mathbf{Z}} \mathbf{Q} = \begin{cases} 0 & , \quad m \text{ odd} \\ \bigoplus 1(-\nu) & , \quad m = 2\nu \,,\ \nu \in \mathbf{Z}\,, \end{cases}$$

for the realization for absolute Hodge cycles. In particular, if X *is a linear variety over* $\mathbf{C}$, *the corresponding statement holds for the Hodge structure on* $H_a(X(\mathbf{C}), \mathbf{Q})$.

(ii) The Hodge and Tate Conjectures are true for linear varieties.

(iii) If k *is a finite field and* X *is a linear variety over* k, *then*

$$r' \otimes \mathbf{Q}_\ell : H_a^{\mathcal{M}}(X, \mathbf{Q}(b)) \otimes \mathbf{Q}_\ell \xrightarrow{\sim} H_a^{\acute{e}t}(\bar{X}, \mathbf{Q}_\ell(b))^{G_k}\,.$$

Remark 10.6.1 Related to the construction of a good category of mixed motives $\mathcal{MM}$, for example over a number field k, such that one should have (among other things)

$$H^i_{\mathcal{M}}(X, \mathbf{Q}(j)) = H^i_{\mathcal{MM}}(X, j) \otimes \mathbf{Q} \quad i, j \in \mathbf{Z}$$

with

$$H^i_{\mathcal{MM}}(X, j) = \mathrm{Ext}^i_{\mathcal{MM}}(\mathbf{Z}(0), \mathbf{Z}(j)) = \mathbf{H}^i(X, \mathbf{Z}(j)^\bullet_{\mathcal{M}})$$

for a suitable complex $\mathbf{Z}(\star)^\bullet_{\mathcal{M}}$, cf. Conjecture 4.8.1, one should modify the étale cohomology to obtain reasonable ℓ-adic realizations. This is done in [Ja2] and [Ja3]. The result is a Poincaré duality theory $(\tilde{H}^\bullet_{\acute{e}t}, \tilde{H}^{\acute{e}t}_\bullet)$. Then, closely related to Beilinson's first conjecture, one can state conjectures relating this last Poincaré duality theory to continuous group cohomology (a little bit modified). The upshot is that these conjectures are true for linear varieties over the number field k. A similar remark applies to finite fields and global function fields.

Bibliography

[Ar] S. Arakelov. *An intersection theory for divisors on an arithmetic surface.* Izv. Akad. Nauk. SSSR **38** (1974), pp. 1179-1192 .

[Be1] A. Beilinson. *Higher regulators and values of L-functions.* J. Sov. Math. **30** (1985), pp. 2036-2070.

[Be2] A. Beilinson. *Higher regulators of modular curves.* In: Contemp. Math. **55** Part I, AMS (1985), pp. 1-34.

[Be3] A. Beilinson. *Notes on absolute Hodge cohomology.* In: Contemp. Math. **55** Part I, AMS (1985), pp. 35-68.

[Be4] A. Beilinson. *Height pairings for algebraic cycles.* Lecture Notes in Math. **1289** (1987), Springer-Verlag, pp. 1-26.

[BBD] A. Beilinson, J. Bernstein, P. Deligne. *Faisceaux pervers.* Astérisque **100**, SMF (1982).

[BMS] A. Beilinson, R. MacPherson, V. Schekhtman. *Notes on motivic cohomology.* Duke Math. J. **54** (1987), pp. 679-710.

[BFM] P. Baum, W. Fulton, R. MacPherson. *Riemann-Roch for singular varieties.* Publ. Math. IHES **45** (1975), pp. 101-146.

[Bla] D. Blasius. *On the critical values of Hecke L-series.* Invent. Math. **124** (1986), pp. 23-63.

[Bl1] S. Bloch. *A note on height pairings, Tamagawa numbers, and the Birch & Swinnerton-Dyer conjecture.* Invent. Math. **58** (1980), pp. 65-76.

[Bl2] S. Bloch. *Algebraic cycles and the values of L-functions.* J. Reine Angew. Math. **350** (1984), pp. 94-108.

[Bl3] S. Bloch. *Algebraic cycles and higher K-theory.* Adv. Math. **61** (1986), pp. 267-304.

[BG] S. Bloch, D. Grayson. *K_2 and L-functions of elliptic curves: computer calculations.* In: Contemp. Math. **55** Part I (1986), pp. 79-88.

[BO] S. Bloch, A. Ogus. *Gersten's conjecture and the homology of schemes.* Ann. Sc. ENS (4) **7** (1974), pp. 181-202.

[Bo1] A. Borel. *Stable real cohomology of arithmetic groups.* Ann. Sc. ENS (4) **1** (1975), pp. 235-272.

[Bo2] A. Borel. *Cohomologie de SL_2 et valeurs de fonctions zeta aux points entiers.* Ann. Sc. Norm. Pisa (1976), pp. 613-636.

[Co] H. Cohn. *Advanced Number Theory.* Dover Publications (1980).

[Cox] H. Coxeter. *The functions of Schläfli and Lobatschefsky.* Quat. J. Math. **6** (1935), pp. 13-29.

[CW] J. Coates, A. Wiles. *On the conjecture of Birch & Swinnerton-Dyer.* Invent. Math. **39** (1977), pp. 223-251.

[De1] P. Deligne. *Théorie de Hodge III.* Publ. Math. IHES **44** (1974), pp. 5-78.

[De2] P. Deligne. *Valeurs de fonctions L et périodes d'intégrales.* In: Proc. Symp. Pure Math. **33** Part II, AMS (1979), pp. 313-346.

[De3] P. Deligne. *La conjecture de Weil II.* Publ. Math. IHES **52** (1980), pp. 137-252.

[De4] P. Deligne. *Le groupe fondamental de la droite projective moins trois points.* In: Galois Groups over **Q**, Edited by Y. Ihara, K. Ribet, J.-P. Serre, Springer-Verlag (1989), pp. 79-297.

[DMOS] P. Deligne, J. Milne, A. Ogus, K-y. Shih. *Hodge cycles, Motives, and Shimura Varieties.* Lecture Notes in Math. **900** (1982), Springer-Verlag.

[Den] C. Deninger. *Higher regulators of elliptic curves with complex multiplication.* In: Sém. Th. Nombres Paris 1986/1987, Birkhäuser (1988), pp. 111-128.

[EV] H. Esnault, E. Vieweg. *Deligne-Beilinson cohomology.* In: [RSS], Academic Press (1988), pp. 43-91.

[Fa] G. Faltings. *Endlichkeitssätze für abelsche Varietäten über Zahlkörper.* Invent. Math. **73** (1983), pp. 349-366.

[vdG] G. van der Geer. *Hilbert Modular Surfaces.* Springer-Verlag (1988).

[Ge1] S. Gelbart. *Automorphic forms on adèle groups.* Annals of Math. Studies **83**, Princeton Univ. Press (1975).

[Ge2] S. Gelbart. *Automorphic Forms and Artin's Conjecture.* Lecture Notes in Math. **627** (1977), Springer-Verlag, pp. 241-276.

[Gi1] H. Gillet. *Riemann-Roch theorems for higher algebraic K-theory.* Adv. Math. **40** (1981), pp. 203-289.

[Gi2] H. Gillet. *An introduction to higher dimensional Arakelov theory.* In: Contemp. Math. **67**, AMS (1987), pp. 209-228.

[GS] H. Gillet, C. Soulé. *Intersection sur les variétés d'Arakelov.* C.R. Acad. Sc. Paris **299** (1984), pp. 563-566.

[Go] D. Goldfeld. *Gauß's Class Number Problem for Imaginary Quadratic Fields.* Bull. AMS **13**, 1 (1985), pp. 23-37.

[Gon] A. Goncharov. *The Classical Trilogarithm, Algebraic K-Theory of Fields, and Dedekind Zeta-Functions.* Bull. AMS **24**, 1 (1991), pp. 155-162.

[Gr1] A. Grothendieck. *On the de Rham cohomology of algebraic varieties.* Publ. Math. IHES **29** (1966), pp. 95-103.

[Gr2] A. Grothendieck. *Hodge's general conjecture is false for trivial reasons.* Topology **8** (1969), pp. 299-303.

[Gro] B. Gross. *Arithmetic on Elliptic Curves with Complex Multiplication.* Lecture Notes in Math. **776** (1980), Springer-Verlag.

[GZ] B. Gross, D. Zagier. *Heegner points and derivatives of L-series.* Invent. Math. **84** (1986), pp. 225-320.

[Ha] H. Hasse. *Über die Klassenzahl abelscher Zahlkörper.* Nachdruck, Springer-Verlag (1985).

[Hi] F. Hirzebruch. *Topological Methods in Algebraic Geometry.* Springer-Verlag (1966).

[HS] G. Harder, N. Schappacher. *Special values of Hecke L-functions and abelian integrals.* Lecture Notes in Math. **1111** (1985), Springer-Verlag, pp. 17-49.

[HW] M. Heep, U. Weselmann. *Deligne's Conjecture.* In: [RSS], Academic Press (1988), pp. 37-42.

[Ja1] U. Jannsen. *Deligne homology, Hodge $\mathcal{D}$-conjecture and motives.* In: [RSS], Academic Press (1988), pp. 305-372.

[Ja2] U. Jannsen. *Mixed Motives and Algebraic K-Theory.* Lecture Notes in Math. **1400** (1990), Springer-Verlag.

[Ja3] U. Jannsen. *Continuous étale cohomology.* Math. Ann. **280** (1988), pp. 207-245.

[Jo] J.-P. Jouanolou. *Une suite exacte de Mayer-Vietoris en K-théorie algébrique.* Lecture Notes in Math. **341** (1973), Springer-Verlag, pp. 293-316.

[Ka] M. Karoubi. *K-theory, An Introduction.* Springer-Verlag (1978).

[KM] N. Katz, B. Mazur. *Arithmetic moduli of elliptic curves.* Annals of Math. Studies **108**, Princeton Univ. Press (1985).

[KaM] N. Katz, W. Messing. *Some consequences of the Riemann Hypothesis for Varieties over Finite Fields.* Invent. Math. **23** (1974), pp. 73-77.

[Kl] S. Kleiman. *Motives.* In: Proc. 5th Nordic Summer School, Oslo 1970, Wolters-Noordhoff (1972), pp. 53-96.

[Ko] N. Koblitz. *Introduction to Elliptic Curves and Modular Forms.* Springer-Verlag (1984).

[La] S. Lang. *Elliptic Functions.* Addison-Wesley Publishing Company, Inc. (1973).

[Lo] J. Loday. *K-théorie algébrique et représentations de groupes.* Ann. Sc. ENS (4) **9** (1976), pp. 309-377.

[Ma] J. Martinet. *Character theory and Artin L-functions.* In: A. Frölich. *Algebraic Number Fields.* Academic Press (1977), pp. 1-87.

[Man] Yu. Manin. *Correspondences, motives and monoidal transformations.* Math. Sbornik **77** AMS Translations (1970), pp. 439-470.

[Me] J. Mestre. *Formules Explicites et Minorations de Conducteurs de Variétés Algébriques.* Comp. Math. **58** (1986), pp. 209-232.

[Mi] J. Milnor. *Introduction to algebraic K-theory.* Annals of Math. Studies **72**, Princeton Univ. Press (1971).

[Mu] D. Mumford. *Rational equivalence of 0-cycles on surfaces.* J. Math. Kyoto Univ. **9** (1968), pp. 195-204.

[Ne1] J. Neukirch. *Class field theory.* Springer-Verlag (1986).

[Ne2] J. Neukirch. *The Beilinson conjecture for algebraic number fields.* In: [RSS], Academic Press (1988), pp. 193-247.

[No] R. Noguès. *Théorème de Fermat, son histoire.* A. Blanchard (1966).

[Qu] D. Quillen. *Higher algebraic K-theory I.* Lecture Notes in Math. **341** (1973), Springer-Verlag, pp. 85-147.

[Ra1] D. Ramakrishnan. *Analogs of the Bloch-Wigner function for higher polylogarithms.* In: Contemp. Math. **55** Part I, AMS (1986), pp. 371-376.

[Ra2] D. Ramakrishnan. *Arithmetic of Hilbert-Blumenthal surfaces.* CMS Proceedings Vol.**7** (1987), pp. 285-370.

[Ra3] D. Ramakrishnan. *Regulators, algebraic cyles, and values of L-functions.* In: Contemp. Math. **83**, AMS (1989), pp. 183-310.

[Rap] M. Rapoport. *Comparison of the regulators of Beilinson and of Borel.* In: [RSS], Academic Press (1988), pp. 169-192.

[RSS] M. Rapoport, N. Schappacher, P. Schneider (eds). *Beilinson's conjectures on special values of L-functions.* Academic Press (1988)

[Ru] K. Rubin. *Tate–Shafarevich Groups of Elliptic Curves with Complex Multiplication.* In: Algebraic Number Theory (in honor of K.Iwasawa), Edited by J. Coates, R. Greenberg, B. Mazur and I. Satake, Academic Press (1989), pp. 409-419.

[Sa] N. Saavedra Rivano. *Catégories Tannakiennes.* Lecture Notes in Math. **265** (1972), Springer-Verlag.

[Sc] P. Schneider. *Introduction to the Beilinson conjectures.* In: [RSS], Academic Press (1988), pp. 1-35.

[Sch] N. Schappacher. *Periods of Hecke characters.* Lecture Notes in Math. **1301** (1988), Springer-Verlag.

[Se1] J.-P. Serre. *Facteurs locaux des fonctions zêta des variétés algébriques (définitions et conjectures).* Sém. Delange-Pisot-Poitou 1969/1970 (1970), exp. 19.

[Se2] J.-P. Serre. *Modular functions of weight one and Galois representations.* In: Algebraic Number Fields, ed. A. Fröhlich, Academic Press (1977), pp. 193-268.

[Sh] G. Shimura. *Introduction to the Arithmetic Theory of Automorphic Functions.* Iwanami Shoten and Princeton University Press (1971).

[Si] J. Silverman. *The arithmetic of elliptic curves.* Springer-Verlag (1986).

[So] C. Soulé. *Opérations en K-théorie algébrique.* Can. J. Math. **37** (1985), pp. 488-550.

[ST] J.-P. Serre, J. Tate. *Good reduction of abelian varieties.* Ann. of Math. **88** (1968), pp. 492-517.

[Su] A. Suslin. *Algebraic K-theory of fields.* In: Proceedings of the International Congress of Mathematicians 1986, AMS (1987).

[Ta1] J. Tate. *Algebraic cycles and poles of zeta functions.* In: Arithmetic Algebraic Geometry, ed. O.F.G. Schilling, Harper & Row (1965), pp. 93-111.

[Ta2] J. Tate. *Endomorphisms of abelian varieties over finite fields.* Invent. Math. **2** (1966), pp. 134-144.

[Ta3] J. Tate. *The arithmetic of elliptic curves.* Invent. Math. **23** (1974), pp. 179-206.

[Ta4] J. Tate. *Algorithm for determining the type of a singular fiber in an elliptic pencil.* Lecture Notes in Math. **476** (1975), Springer-Verlag, pp. 33-52.

[Ta5] J. Tate. *Relations between K_2 and Galois cohomology.* Invent. Math. **36** (1976), pp. 257-274.

[Ta6] J. Tate. *Les Conjectures de Stark sur les Fonctions L d'Artin en $s = 0$.* Birkhäuser (1984).

[TZ] H. Tschöpe, H. Zimmer. *Computation of the Néron-Tate height on elliptic curves.* Math. Comp. (48) **177** (1987), pp. 351-370.

[Ve] J.-L. Verdier. *Catégories Dérivées, Quelques résultats (État 0).* SGA4$\frac{1}{2}$, Lecture Notes in Math. **569** (1977), pp. 262-311.

[Wa1] L. Washington. *Introduction to Cyclotomic Fields.* Springer-Verlag (1982).

[Wa2] L. Washington. *Number Fields and Elliptic Curves.* In: Number Theory and Applications (NATO ASI Series), Edited by Richard A. Mollin, Kluwer Academic Publishers (1989), pp. 245-278.

[We] A. Weil. *Adèles and Algebraic Groups.* IAS, Princeton (1961)

[Za1] D. Zagier. *Hyperbolic manifolds and special values of Dedekind zeta-functions.* Invent. Math. **83** (1986), pp. 285-301.

[Za2] D. Zagier. *Polylogarithms, Dedekind Zeta Functions, and the Algebraic K-Theory of Fields.* In: Arithmetic Algebraic Geometry, Edited by G. van der Geer, F. Oort, J. Steenbrink, Progress in Mathematics **89**, Birkhäuser (1990), pp. 391-444.

Index

Aspects of Mathematics

Edited by Klas Diederich

*A publication of the Max-Planck-Institut für Mathematik, Bonn

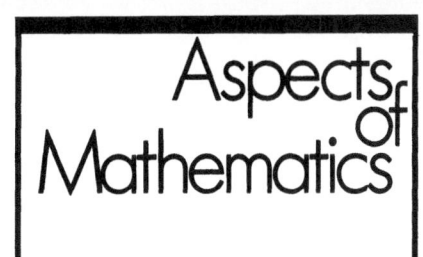

Complex Analysis

Dedicated to H. Grauert

Proceedings of the International Workshop 1990
Edited by Klas Diederich (Ed.)

*1991. X, 341 pp. (Aspects of Mathematics, Vol. E 17;
ed. by Klas Diederich) Hardcover.
ISBN 3-528-06413-7*

This volume contains the Proceedings of the International Workshop "Complex Analysis", which was held from February 12-16, 1990, in Wuppertal (Germany) in honour of H. Grauert, one of the most creative mathematicians in Complex-Analysis of this century. In complete accordance with the width of the work of Grauert the book contains research notes and longer articles of many important mathematicians from all areas of Complex Analysis (Altogether there are 49 articles in the volume). Some of the main subjects are: Cauchy-Riemann Equations with estimates, q-convexity, CR structures, deformation theory, envelopes of holomorpy, function algebras, complex group actions, Hodge theory, instantons, Kähler geometry, Lefschetz theorems, holomorphic mappings, Nevanlinna theory, complex singularities, twistor theory, uniformization.

Vieweg Publishing · P. O. Box 58 29 · D-6200 Wiesbaden / FRG

vieweg